Introduct
Physical

Part 2

A

Introductory Physical Science

Part 2

by **D. C. FIRTH**
Senior Science Master
Bristol Grammar School

and **K. W. LYON**
Senior Physics Master
Bristol Grammar School

CAMBRIDGE
at the University Press 1971

Published by the Syndics of the Cambridge University Press
Bentley House, 200 Euston Road, London N.W.1
American Branch: 32 East 57th Street, New York, N.Y.10022

Library of Congress Catalogue Card Number: 70–142127

ISBNs: 0 521 07644 7 Part 1
0 521 08037 1 Part 2

Printed in Great Britain by
Ebenezer Baylis & Son Ltd, Leicester and London

Contents

Acknowledgements

The authors and publisher would like to thank the following for supplying photographs for use in this book:

Keystone Press Agency Ltd for figures 13.7, 13.24, 13.25, 21.2, 21.8, 23.2, 23.14, 24.14; Kodak Ltd for 13.9, 23.3; Radio Times Hulton Picture Library for 13.11, 21.7; Austrian National Tourist Office for 13.12; National Ski Federation of Great Britain for 13.15; USIS for 13.17, 16.6; British Oxygen Ltd for 13.20, 18.4, 20.8, 20.10, 20.11, 20.13, 21.6, 21.9; Esso Petroleum Ltd for 13.21; the Science Museum for 13.23, 23.40; Associated Electrical Industries Ltd for 14.2; the British Travel Association for 15.3; Fibreglass Ltd for 15.12; National Physical Laboratory of Israel for 15.14; General Electric Company for 15.21; Prestige Ltd for 15.22; Sartorius Balances Ltd for 16.2; Griffin and George Ltd for 19.19, 19.47; Rolls Royce Ltd for 19.26; British Aircraft Corporation for 19.27; British Leyland for 19.40; James Neill and Company for 19.42; National Coal Board for 20.9, 25.1; British Steel Corporation for 20.12; National Aeronautics and Space Administration for 22.1; British Information Services for 23.13; D. G. A. Dyson for 23.20, 23.21; Trinity College for 23.25; Victor Hopkins for 23.27; The Gas Council for 24.1, 24.3, 24.7, 24.8; Imperial Chemical Industries for 24.19; De Beers Consolidated Mines for 25.4; Industrial Diamond Information Bureau for 25.7.

Photographs of experiments are by S. C. Moreton-Prichard; the drawings are by Brian Denyer.

Publisher's Note: *Introductory Physical Science* is published in two parts, both of which contain Chapter 13 'Energy'. A *Teacher's Guide* is published separately.

Chapter 13 **Energy**

You have learned that energy is something that causes things to change, and we have already come across two ways of giving things energy. One was to heat them. We decided that heating things could cause two kinds of change. (What are their names?) Another was to send an electric current through things. We must now think a bit more about energy because, as we said earlier, one of the jobs of a scientist is to change things – or sometimes to stop them changing. Name one domestic appliance that is used to change things and one that is used to stop them from changing.

You may have heard your mother say (perhaps when she was a bit tired) 'Oh, I wish I had your energy!' or 'How full of energy you are!'. What does she mean? Well, that you are active, busy, or work hard. When the scientist uses the words 'energy', and 'work', he doesn't mean quite what we mean when we use them in daily life, but that doesn't mean that there is no connection at all between the scientist's use and ours.

Experiment

You will need a thermometer, a block of wood with a flat surface and big enough to hold in your hand. You will also need a rough wood surface – a board, a plank or a carpenter's bench – but *not* a polished piece of furniture. Lay the thermometer on the board, cover it with a handkerchief to prevent draughts, and leave it for two or three minutes so that it will show the temperature of the board. Record this. Now remove the thermometer, place the block on the board and rub it to and fro a few centimetres, pressing down really hard and rubbing quickly for one minute (time yourself or count 100 rubs). Quickly put the thermometer and handkerchief back as before and watch the reading.

Has the temperature of the board changed? How? Has it got some heat from somewhere? Where has it come from? Do you feel a bit tired? Have you, perhaps, supplied the energy? Was it 'hard work' doing this experiment? Do you yourself feel warmer anywhere after doing this? Could someone with

(a)

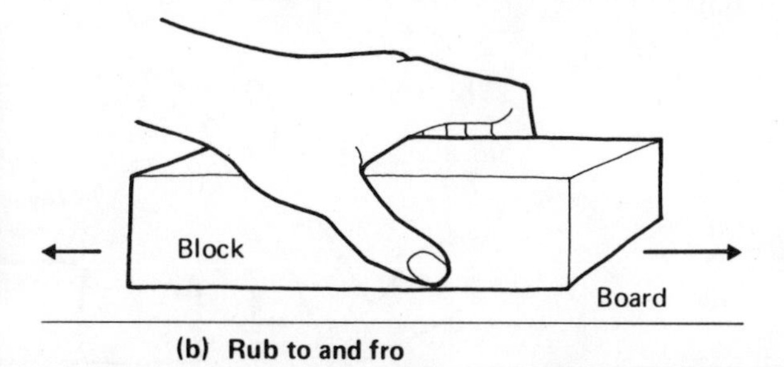

(b) Rub to and fro

Fig. 13.1. This boy is heating the board by rubbing it. The way to do this is shown in diagram (*b*). Have you ever been told to use more 'elbow grease'? What word would a scientist use?

feeble muscles – a very old lady for instance – do this experiment successfully?

You know that in your daily life you don't get money from nowhere. If father gives you pocket money, then what goes into your pocket comes out of his. If he wins money on a football pool it comes from the other people who have lost on that pool. The same is true of energy. If something gets extra energy this has come from somewhere else, it doesn't just happen. So the heat energy that appeared in the warm board was stored up somehow in your muscles and if your muscles hadn't enough of it you couldn't have done the job successfully. After you finished rubbing you may have felt a bit tired or stiff and perhaps warm. You will learn later on that this was because chemical changes had gone on inside your muscles. So what happened was that *chemical* energy inside your body was changed into *heat* energy in the board. We can write this as follows:

Chemical energy in muscles ⟶ Heat energy in board

and we call this an 'energy diagram'.

Fig. 13.2.

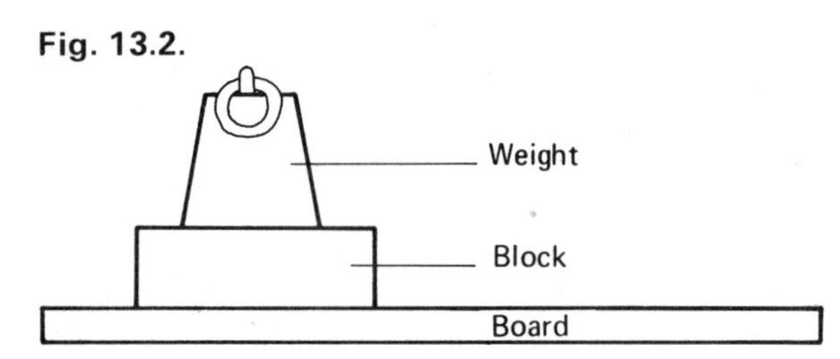

Experiment

Use the same apparatus as before and record the temperature of the board as you did last time. Put the block on the board and then stand a heavy weight on top of it. Do not rub the board. After a minute take off the weight and block and record the temperature of the board again. Has the temperature risen this time?

To warm the board you don't just have to press hard, you have to *move* the block over it as well, and scientists say you have done some *work*. Whenever you move something so that energy changes from one kind to another, work is done. If you want to change the chemical energy in your muscles into some other kind of energy you have to do some work.

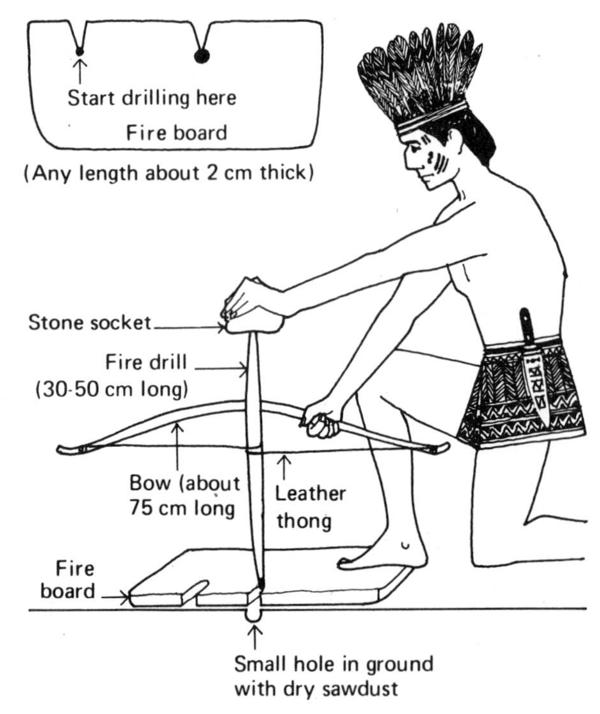

Fig. 13.4. Changing mechanical energy into heat – a Red Indian's fire-making kit. As he moves the bow to and fro it heats the fire board and also wears it down to dust. The dust falls into the little hole (about 2 cm deep) and eventually catches fire. One foot holds down the board. The thong is given one turn round the drill. The drill is better not round but hexagonal. You can try making this at home. You should blow gently on the sawdust when it begins to smoke. Write an energy diagram for this change.

Fig. 13.3.

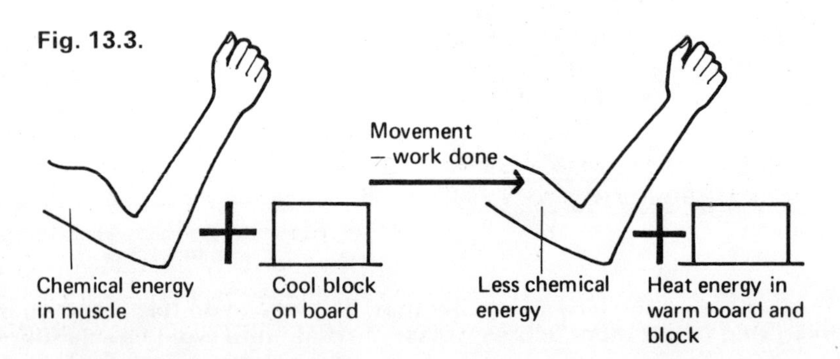

There is really nothing strange in chemicals having energy stored in them. We found earlier (p. 93) that some chemicals gave out heat when you warmed them up a little in a tube and started off a chemical change. For example, wax caught fire if it was heated enough. Think of two more examples. These things had chemical energy in them, and once you started it off by warming the substances, this chemical energy changed to heat energy. You may feel it was cheating because you warmed the tube, but many chemicals will turn their energy into heat without any heating at all.

Experiment

(1) Put a few pieces of zinc into a test tube and pour in copper sulphate solution until the tube is about one-third full. Hold the bottom of the tube. What do you feel? Look carefully at the tube again after a few minutes. What do you see? Has there been a chemical change? Has some heat energy appeared? We can write:

Chemical energy in zinc and copper sulphate ⟶ **Heat energy in liquid**

Fig. 13.5.

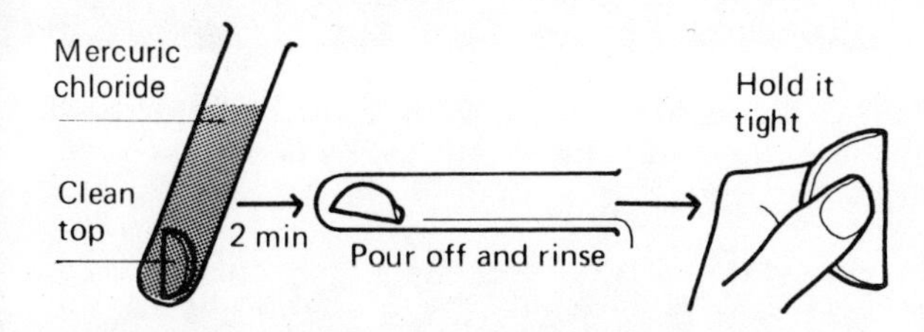

(2) Pour 2 cm of mercuric chloride into a test tube. Put into it a 'silver' metal milk bottle 'top' that you have first washed well with soap and water. After two minutes drain off the mercuric chloride into a sink. Pour water into the tube to wash the top and pour the wash water down the sink. Rinse the top three times like this. Poke the top out of the tube and hold it firmly between your finger and thumb. Keep it there.

Complete the energy diagram for this experiment which is started below:

Chemical energy in metal and moisture in your hand ⟶

Warning: **Mercuric chloride is poisonous – handle carefully and wash your hands when the experiment is over.**

We can now think about three kinds of energy: heat energy, electrical energy and chemical energy. We know that chemical energy can change into heat energy. When you switch on an electric fire or lamp, you have proof that electrical energy can turn into heat energy since you can feel the heat. Can we turn heat energy into electrical energy?

Experiment

Take a piece of copper wire and a piece of some other kind of wire and twist them together at one end. Thread the wires through a piece of asbestos (Fig. 13.6) and connect to an electric meter. Then heat the twisted ends in the flame of a Bunsen burner. What happens? Does the meter's pointer move? Is electrical energy used to move it? When the wires have cooled down examine them again. Have they been through a physical change or a chemical one?

Fig. 13.6.

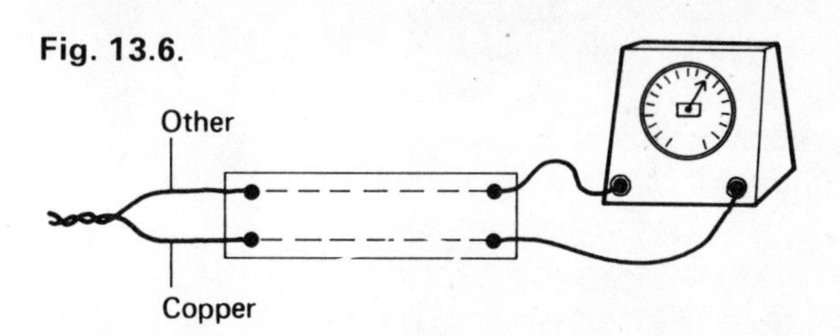

An arrangement like the one you have used is called a *thermocouple*, and it turns heat energy straight into electrical energy.

Something to do at home

If you have an electrical set at home and if you have a meter that will read small currents (a few milli-ampères – what is a milliampère?), you can try various pairs of wires – copper and tin, copper and nichrome (from a broken fire element) or other kinds to see which of them gives the biggest current when you heat them. Record carefully what you find. How can you make sure that each pair of wires is heated the same amount?

Motion energy

You don't always use your muscles to rub things as you did when you heated the bench by rubbing. What else can the chemical energy in your muscles be used to do? You have often thrown or hit a ball when playing rounders, tennis or football, and if you do it hard and often you will feel tired. Some of you may have been unlucky or careless enough to hit a ball through a window or a greenhouse roof. If so, you caused a change in the glass. (Was it a physical or a chemical change?) The ball evidently had some energy because it changed something. You have all seen pictures of vehicles battered and

Fig. 13.7. The ball was hit by G. Stilwell and A. Ashe is about to return it. Why is the ball blurred? What kind of energy has it? What kinds of energy have the man? the racket? Why is not the man also blurred?

buckled in road accidents. They have certainly been very much changed. This happened because before the collision they were *moving*. So moving bodies can change things and must have energy. We shall call this kind of energy *motion* energy. Scientists often use the name *kinetic* energy for this, so if you meet this word you will know what it means.

Experiment

Take a nail, a piece of wood and an iron retort stand. Try to *press* the nail into the wood with the stand (Fig. 13.10 (*a*)). Is this a good way to drive the nail into the wood? Now use the retort stand as a hammer – or use a real hammer – and drive the nail in (Fig. 13.10 (*b*)). The hammer moves the nail into the wood and does work as it goes in. What kind of energy has the hammer just before it hits the nail? How did this energy get into the hammer? Where did it come from? Could a weak person have hammered the nail in? Did you move anything while you were hammering? Did you do any work?

Fig. 13.8. An unwanted change. These two vehicles have been terribly changed because they collided. What kind of energy had they before they met? Into what kind of energy was most of it changed? It is because moving vehicles have so much energy that there are strict rules to be observed by those who drive them.

Fig. 13.9. Making use of Nature's energy. One of the oldest ways to tap natural energy is to use a windmill. What kind of energy does wind have? What use was made of the energy in a mill like this? What other kinds of mill were built in olden times to enable man to use Nature's energy?

Fig. 13.10.

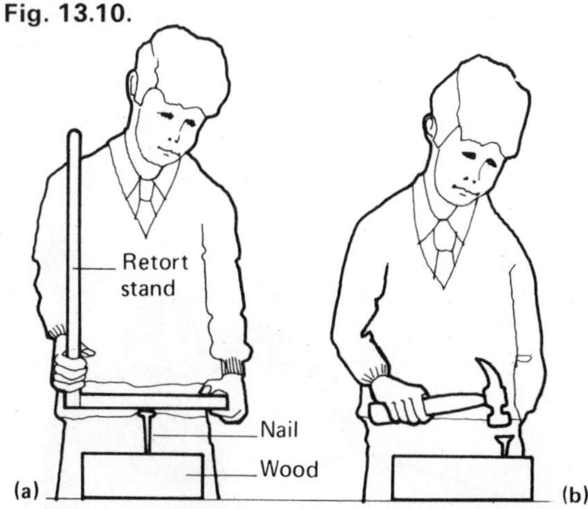

Hammer another nail half-way into a piece of wood, then quickly pull it out and hold the nail against the back of your hand. What do you notice? Has the nail got some energy from the hammer? Did the wood get any?

The energy diagram for this little experiment is written:

Chemical energy in muscles ⟶ Motion energy in hammer ⟶ Heat energy in nail and wood

Fig. 13.11. Cooling the work. When metal is worked on a lathe much of the motion energy is converted into heat. Cooling liquid is poured over the metal while the machine is working.

Fig. 13.12. Climbing in the Dolomites. What kind of energy does the climber have when he is high up like this? Where did this energy come from originally? What kind of energy is stored in the rope?

Place energy

We have just discovered another kind of energy– motion energy, so now we know about four kinds of energy. What are they? Are there any more? You don't always use your muscles to throw things– sometimes you lift them, put them on shelves or tables. When you climb a hill, a rope or a ladder your leg and arm muscles lift you to the top. You feel tired when you have done these things, and since you have moved something you have done some work. Some of the chemical energy in your muscles has been changed, but where has it gone? To find this out we will do two little experiments.

Experiment

You will need a piece of string, a small dynamo – a cycle dynamo will do – a flash-lamp bulb and some connecting wire. The dynamo must be firmly held or fixed to the bench and then joined to the bulb with connecting wires (Fig. 13.13). Wind the string round the spindle of the dynamo and then pull it steadily and firmly. Watch the bulb.

Fig. 13.13.

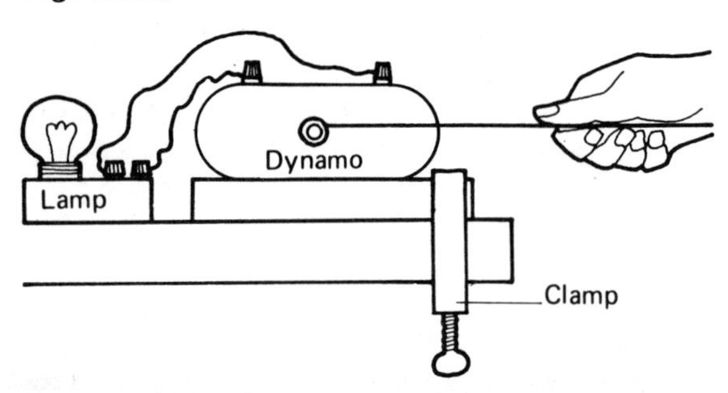

Does the bulb get hot and light up? If so its heat energy must have come from somewhere. It seems to have come from the dynamo, so what kind of energy did the bulb turn into heat? But where did the dynamo get its energy from? Did you do any work? What moved in this experiment? Will the lamp light if no one moves the dynamo?

In this experiment we have used chemical energy in our muscles to produce electrical energy, and then this electrical energy has been changed into heat in the bulb. We can write this energy diagram:

Chemical energy in muscles → Electrical energy in dynamo → Heat energy in bulb

Experiment

Now arrange things a little differently as in Fig. 13.14. Tie the weight to one end of the string and wind the other round the spindle as before. Use a pulley wheel if this is more convenient. Release the weight gently so that it falls to the ground. Does the bulb light? What is turning the dynamo and doing the work this time?

Fig. 13.14.

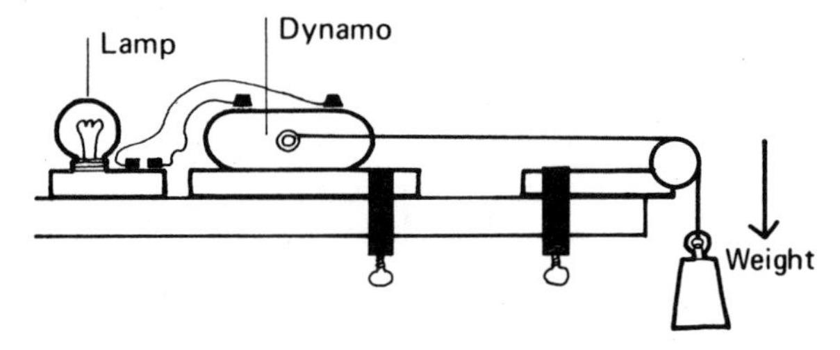

While the weight went down it turned the dynamo and it also picked up speed itself, so it gained some motion energy as well as turning the dynamo. It must have had quite a lot of energy stored in it, but how?

When the weight is above the ground it can fall, and so it can pull the string and turn the dynamo, but once it reaches the floor it can't do this any more. So it seems as if a weight high up has more energy than a weight low down. We said earlier that if you lift a heavy thing on to a shelf you have to do some work to get it there and now we have discovered that it has more energy on the shelf than it had before. The chemical energy in your muscles has been changed into this kind of energy. It seems

Fig. 13.15. Skiing. You have to be energetic to enjoy skiing. What kinds of energy has the man in the picture? What energy changes take place while he is climbing or being 'lifted' to the top of the run?

Fig. 13.16. Bodiam Castle, Sussex. The defenders of a castle such as this often carried heavy stones to the tops of the towers. What energy changes occurred when they did this? What kind of energy did the stones get when dropped on the enemy below? What happens to this energy when the stone falls into the water?

Fig. 13.17. The Robert Moses Niagara Power Plant. In the building in the front of the picture the place energy of the water behind the building is changed into electrical energy. At off-peak times, when this energy is not needed by factories it is passed to the other plant at the back and used to pump water up into the lake behind. This provides extra place energy for use during peak hours.

that sometimes things can have energy simply because of where they are. We can call this kind of energy *'place energy'* and we can write this energy diagram:

Chemical		Place		Electrical		Heat
energy in	→	energy of	→	energy in	→	energy
muscles		weight		dynamo		in bulb

Place energy is another kind to add to our list. Write down the various kinds of energy we have found so far – there are five.

Fig. 13.18.

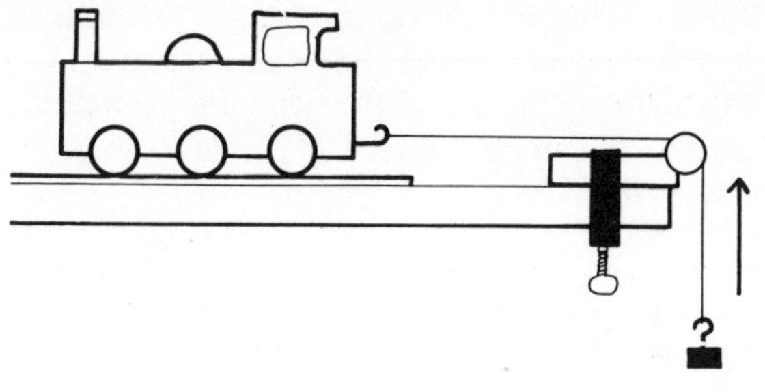

Experiment

Wind up the little clockwork engine, put it on the track and hook on to it the thread attached to the weight. Take off the brake and see if the engine will pull the weight up – if not try a smaller weight. You know that the engine can pull carriages, so it can lift a weight if the weight isn't too big. What kind of energy has the engine when it is moving? What kind has the weight after it has been lifted?

Before you started the engine, this energy must have been somewhere inside it. Your muscles put the energy there because you wound up the engine, but where was the energy until you pressed the starting lever? It isn't electrical energy because there is no electrical gear in a clockwork motor, and it can't be chemical energy because the engine doesn't use up any chemicals. It isn't place energy because the engine didn't run downhill while lifting the weight.

Experiment

Take the outside off the engine so that you can see what happens when you wind it up and when it unwinds. If you can't get the outside off examine an old clock instead. Do you have to do work to turn the winder? Does the shape of the spring change as you wind it up? Is this a physical or a chemical change? Does the spring go back to its original shape as it unwinds?

When the spring unwinds it turns the wheels of the engine and moves the train or lifts the weight. While it does this it changes its shape. So it seems as though a thing can be given energy if we make it change its shape. We shall call this kind of energy *shape* energy. For our last experiment we can draw this energy diagram:

Chemical energy in muscles → Shape energy of spring → Place energy of weight and Motion energy of engine

Many things besides springs can have shape energy. If you have swung on the branch of a tree you will have watched it or felt it bend under your weight, and you'll have seen it spring back when you let go. When you pulled it down you changed the shape of the tree a bit and gave it some shape energy. If you watch closely the bar of the swing in a recreation ground you can see it change shape a bit as someone swings on it.

Fig. 13.19.

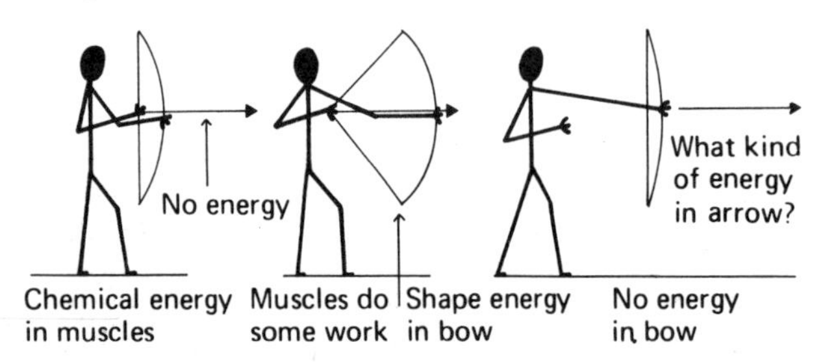

If when you were younger you played with bows and arrows you changed chemical muscle energy into shape energy when you bent the bow. What kind of energy does the arrow have after you've fired it? What happens to this energy eventually?

Experiment

Blow up a rubber balloon and hold the end so that the air doesn't escape. Did you feel as if you were doing some work while this was happening? Has the air in the balloon the same volume as it had when it was inside you? What kind of energy did you give the air, then? Release the balloon. What happens to it? What kind of energy has it as it goes through the air? Write an energy diagram for this experiment similar to those for the earlier experiments.

Light energy

All of you, I am sure, have watched somebody take a photograph, and some of you have probably taken one yourself. You will learn more about cameras later on, but you know that the photographer puts a film into the camera, then he points it at the view he wants to take – let us say the Great Pyramid in Egypt. Then he presses a catch and this opens a shutter for a short time and lets the light in, and the job's done. Something funny happens to the film. After the photo is taken, it is sent to be developed and you get back a picture.

After the shutter has been opened the film is changed. Is it a physical or a chemical change? Does the film 'change back' while you are using up the rest of the roll? The film hadn't got the Great Pyramid on it when you bought it and the Great Pyramid wouldn't get onto it if you didn't let the light in at the right time. So the light must have caused the change. Since light can change a film in this way it must be another kind of energy.

Fig. 13.20. A lighthouse. For many years men have used light energy to carry warnings of rocky coasts to ships at sea. What kind of energy can be used when it is foggy?

Experiment

Pour some silver nitrate solution into a test tube until it is about one-quarter full. Pour in dilute hydrochloric acid until the tube is half full. What do you see? Leave the tube on a windowsill for about half an hour. Does what is in the tube look different in any way?

The white precipitate you saw was silver chloride. What elements are there in it? The precipitate goes darker because light changes this into silver.

We have now met with seven different kinds of energy. There are still more kinds. For example, you may have heard your windows rattle when a crack of thunder sounded rather near. If they rattled they must have got some motion energy from somewhere and you can guess that this came from the thunder, so *sound* is a kind of energy too. You will come across other kinds of energy as you learn more science. Scientists find it very helpful in their work to think about the energy changes that go on and they are always on the look-out for new kinds of energy.

Fig. 13.21. Using energy in a garage. The lift is worked by compressed air. What kind of energy has compressed air? What kind of energy has the car on the lift? What gives the compressed air its energy? How can the man pour oil up hill?

Questions

(1) When a drill is taken quickly out of a hard piece of wood in which it has just made a hole it usually feels hot. Where did its heat energy come from?

(2) When you pump up a bicycle tyre the air inside the tyre gets shape energy. Some energy from your muscles also changes into another form. Which form is this?

(3) Make energy diagrams to show what energy changes take place when the following things happen:

(*a*) A car starts, climbs a steep hill and stops on the top of it.

(*b*) A car on top of a hill runs down it and the driver stops it at the bottom, using his brakes. What would he notice if he felt the brake drums?

(*c*) The little steam engine (Fig. 13.22 (*a*)) winds up a weight from the floor.

(*d*) The water turbine (Fig. 13.22 (*b*)) turns a dynamo that lights an electric bulb?

(*e*) The elastic of a catapult is stretched and a stone is fired from it.

(4) When the hot bar cooled in the frame (p. 33) it broke the cast-iron rod that we put through it. What kinds of energy did the hot bar have? What kind of energy do you think the cast-iron rod had just before it broke? Where did this go?

(5) When you let off a rocket on Guy Fawkes' night some exciting energy changes take place. The rocket contains chemical energy to begin with. Name as many other kinds of energy as you can that this chemical energy changes into.

Fig. 13.22.

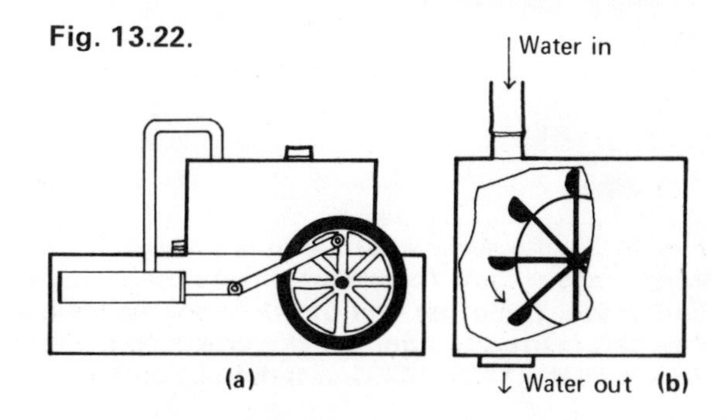

(6) What can happen to curtains after they have been hanging for a year or two at a sunny window? What kind of change do you think this is? What kind of energy do you think has caused this change?

Fig. 13.23. An early steam engine (model). Fuel is burned to produce heat energy and thus change water into steam. The steam enters the cylinder on the right and moves that end of the beam up and down. The other end of the beam moves a rod attached to a gear and so turns the big wheel. The belt is used to connect the wheel to a circular saw. The gear in this picture was invented by James Watt (1736–1819), a famous Scottish engineer. It is called a 'sun and planet' gear. Why?

(7) Who designed the first steam locomotive? The first steam boat?
(8) What is a beam engine?
(9) Where are Boulder Dam? the Aswan High Dam?
(10) Find out what you can about the scientists Joule, Trevithick, Watt.

Summary

Energy makes things change. The changes may be chemical or physical.

There are many kinds of energy. We have talked about chemical, heat, motion, place, shape, electrical, sound and light energy.

You never get energy for nothing. Energy can

only change from one kind to another.

Work is done when energy changes from one kind to another and something moves as well.

Fig. 13.24. This rocket is taller than St Paul's Cathedral. It started Aldrin, Armstrong and Collins on their way to the moon, giving them an enormous amount of motion energy. What kind of energy was changed into this motion energy? Why are people not allowed to go near the rocket when it is fired?

Fig. 13.25. Making use of chemical energy. After the new bridge was finished the old one was no longer needed and would soon have become dangerous. The chemical energy stored up in explosives was used by engineers to break up the unwanted bridge. Into what kinds of energy was the chemical energy changed? Sometimes explosions take place when they are *not* wanted.

Measuring energy

When we talked about lengths and masses, times and temperatures, we went on to say how they were measured, because, as we said then, you can't do any really good job without measurement. Of course, scientists have to measure energy too, but this is difficult because there are so many different kinds of energy. These can all be changed into one another, as we have seen, but some are easier to measure than others. The easiest to measure is place energy.

Fig. 13.26.

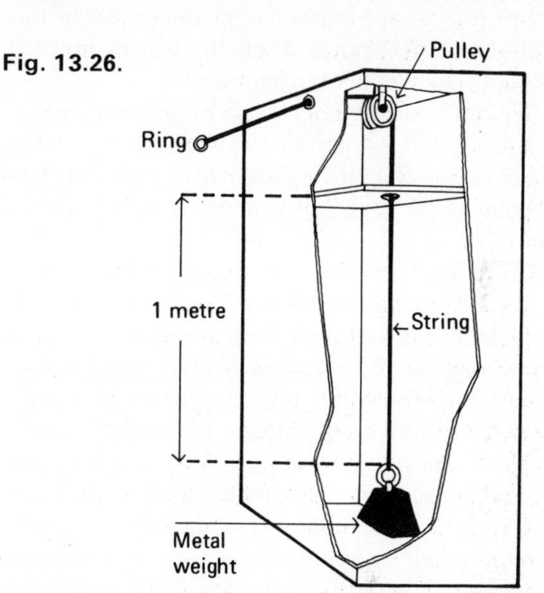

To lift something up from the floor you have to exert a force on it. This is because the earth's gravity pulls it downwards and it can't move up unless you exert a force upwards that is just as big. Figure 13.26 shows a force box. It contains a lump of metal that is attracted by the earth with a force of one newton. Tied to the top of the metal is a string that goes over a pulley and has a ring on the end. When you pull the string the metal rises but it is stopped from going too far by a wooden bar. The bar is arranged so that the metal can rise just 1 metre.

Suppose you pull the metal up until it has risen 1 metre. It now has some place energy that it didn't have before. This amount of extra place energy we call one *joule*. Where has the extra place energy come from? The force you pulled with was 1 newton – or just a tiny bit more. You pulled with this force for a distance of 1 metre. When you use a force of 1 newton and move something 1 metre with it, then 1 joule of energy always changes from one kind of energy to another, and you have done 1 joule of work.

Of course you don't always give things place energy when you do work. When you rubbed the bench with the wood block earlier on you did work. Into what kind of energy was your muscle energy changed then? When you hit a tennis ball you give it motion energy, for instance.

Suppose that after pulling up the weight in the force box you let go the ring when the metal is just against the bar. It would fall. As it falls its place energy changes into motion energy. How much of this kind of energy will it have just before it hits the floor? Into what kind of energy does most of this motion energy change when the weight hits the ground? (Remember the hammer!)

We can measure motion energy and heat energy in joules too. It is not so easy to work out how many joules of energy a moving football or a running boy or a racing car have, but you will learn later that it can be done.

If you had been able to lift the weight *two* metres instead of one, you would have done 1 joule of work in lifting it the first metre and another in lifting it the second, so you would have done 2 joules altogether. Suppose you had lifted *two* weights of this size 1 metre, how many joules would you have done?

The simple rule for finding how much energy you have changed from one kind to another when you pull something is: multiply the number of newtons force you pull it with by the number of metres you move it, and this gives the answer in joules. But you must pull it straight in the direction you want and not at an angle. So we can say:

Energy changed = force × distance you move a thing with it in the direction the force pulls = work you do.

Fig. 13.27.

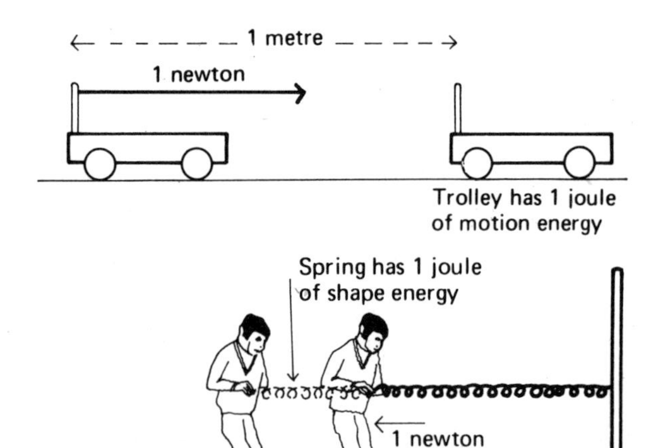

Fig. 13.28.

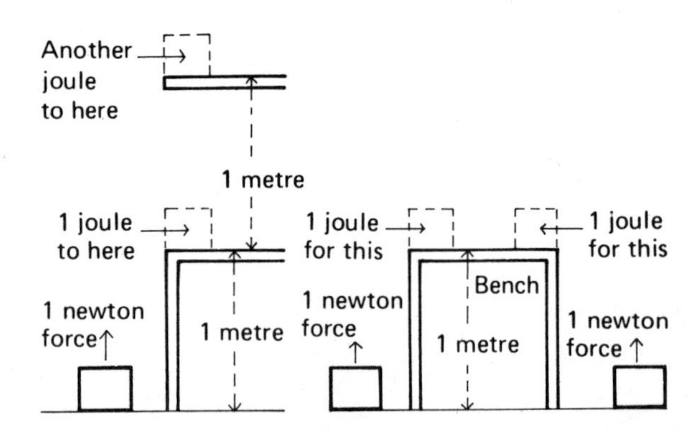

Questions

(1) You pull a trolley with a force of 10 newtons and drag it 100 metres along the road. How much work do you do?

(2) A boy pulls a catapult elastic with a force of 12 newtons, and the stone is drawn back 25 cm. How many joules of energy are stored in the elastic? What kind of energy is it? How much work did the boy do? What happens to the energy when the boy lets go of the elastic?

(3) For every stone you weigh the earth pulls on you with a force of about 62 newtons. If you know your weight, you can work out how many joules of place energy you gain when you go upstairs to bed. An ordinary room is about 2 m high.

(4) A British Rail diesel loco can pull a train with a force of about 150000 newtons. At 144 km/h how far does it travel in 1 second? How many joules of work does it do every second? What changes in energy take place while the engine is working?

Chapter 14 Heat Energy

We have done a number of experiments in which we have supplied different things with energy and studied how they change. We have learned that sometimes the change is a *physical* one and sometimes a *chemical* one. (How do you try to decide whether a change is physical or chemical?) We also know that when some chemical changes take place a lot of heat energy is produced – for example when coal or oil burn. Apparently energy is stored in these substances as chemical energy and comes out of them when they burn. We found that when we heated an iron bar we could store in it enough shape energy to break a cast-iron plug. We are now going to learn more about heat energy and to try to measure it for ourselves.

Experiment: *To study the effect of heating water steadily*

Fig. 14.1.

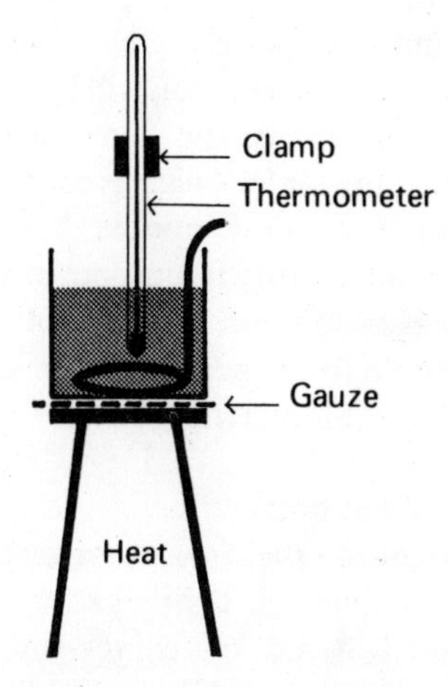

(A) Stand an empty tin or beaker on your tripod. Arrange the stand so that your thermometer bulb is about two inches above the base of the tin and so that you can read it easily all the way up.

Take the stand away again and remove the tin.

Light your Bunsen burner and adjust it to give a small non-luminous flame. (How?) Put the burner under the tripod and gauze and leave it there. It should not now be touched until the experiment is over.

Fill your tin about two-thirds full of water and stand it on the bench. Stir it carefully and record the temperature of the water. Examine your thermometer so that you will be able later to read it quickly. In your book make a table like this:

Time (minutes) 0 $\frac{1}{2}$ 1 $1\frac{1}{2}$ 2 $2\frac{1}{2}$ 3 $3\frac{1}{2}$ up to 5 min.
Temp. (° C)
Rise in $\frac{1}{2}$ min. (° C)

When you are told to do so, put the tin on the tripod and put the thermometer and stand back in their old position. Take care not to bang the thermometer against the tin as it breaks easily. Read the thermometer every half minute and record the temperature in your note-book. Remember to try to read between the divisions. You will have to read the thermometer quickly so it is a good idea to be looking at it before it is really time to read it. There may be a clock in the lab. for you to work by, or your teacher may call out the time every half minute, or you may use your own watch – you will be told what to do. While the water is being heated you must stir it steadily. When you have finished turn out your burner and look at the thermometer from time to time. Put your hand *near* the tin. What do you notice?

(B) You may be supplied with a small electric immersion heater instead of a Bunsen burner. If so, stand the heater in your tin and pour in enough water to cover it completely. Pour this water into your measuring cylinder and add a little more to bring the

Fig. 14.2. A domestic immersion heater. Electric energy is changed to heat energy in the heater and used to warm the water in the cylinder. What is done to keep the heat in the cylinder?

volume up to a convenient one (70, 80, 100, 120 cm^3 for example). Then pour the water back into the tin, put the thermometer into it and wait for at least two minutes so that the temperature gets steady. When you are told you will switch on your heater and record the temperatures as described above.

Questions

(1) Do you think that the burner/heater supplies the same amount of heat energy in each half minute? Is there any reason why it should supply more heat energy in one half minute than in the next?
(2) Does the temperature rise by the same amount in each half minute for the first few times? all the time?
(3) What happens to the temperature after you turn out the flame (or switch off the heater). When you put your hand near the tin does it feel warm? Is the tin then losing energy?
(4) When you were heating the tin, do you think that all the heat energy from the flame/heater stayed in the water?

In doing this experiment you have to read a thermometer rather hurriedly so that your readings may not be quite right. Also if there is a draught it may cool the tin more at one time than another. But if you have been very careful you will have found that the energy supplied by the flame/heater in two minutes warms the water twice as many degrees as the energy supplied in one minute.

It would be interesting to find out what happens when you heat a different quantity of water. You may be able to do this experiment yourself at home, if you have a thermometer.

Something to do at home

Switch on one of the top rings of the electric cooker and adjust it to *'low'* – or light a gas ring and adjust it to give a small flame. Whichever you do, leave it for two or three minutes while you make your other preparations. Choose a big saucepan and pour in two bottlefuls of water, measured with an empty milk or other bottle. If you prefer you can use a small saucepan and pour in two cupfuls or jugfuls of

water. Record the temperature of the water. Put the pan on the ring just when the minute hand of your watch is level with a mark on the dial – if you have a seconds hand you can of course use that. Stir the water with a wooden spoon, holding the thermometer in it with one hand and find how long it takes for the temperature to rise 20° C. Record this time. Pour away the water, rinse the pan with cold tap water until it is cold (Why?), then pour in *one* bottleful (or jugful or cup) of water, record its temperature and then proceed as before, noting the time it takes for the temperature to rise 20° C.

You must have known before you did this experiment that a small quantity of water heats up more quickly than a large quantity. But does one bottleful heat up in half the time that two bottlefuls takes? Would you expect it to? If it doesn't, does it take a bit more than half the time or a bit less? Is anything else heated each time as well as the water? When you halve the water do you halve the 'anything else'? Does all the heat energy that goes into the water stay there? What would happen if you turned off the heat and left the saucepan of water on the stove?

Energy units

Scientists have done many experiments like the ones you have just been doing. They have wrapped up the apparatus very carefully to try to stop the energy they put in from being lost again and have taken other precautions. The results have convinced them of two things:

(*a*) it takes twice as much heat energy to warm up two grams of a substance as it does to warm 1 gram of the substance (the same number of degrees of course);

(*b*) to warm anything two degrees takes twice as much heat energy as to warm it one degree.

Just as we measure lengths in units called centimetres and masses in grams so we must have a unit in which to measure heat energy. Now heat energy as you know, is just one of lots of different kinds of energy and so we can measure it in *joules*. Think about the experiment when you pulled the weight up in the box and gave it one joule of place energy. If the weight was allowed to drop back again, this place energy would turn into heat, so it is quite sensible to measure heat energy in joules.

Some scientists prefer to use a different unit called a *calorie*. A calorie is the heat energy needed to warm 1 g of water 1° C. This is a convenient unit because we probably heat more water in our daily life than we do anything else. Think of all those cups of tea! Scientists have measured very carefully how many joules are needed to heat 1 g of water 1° C and they have found that it takes 4·2 joules. So 1 cal = 4·2 joules, which is worth remembering.

Now you wouldn't get much of a cup of tea if you only warmed 1 g of water 1° C – to make tea we need to heat many grammes of water from room temperature to boiling point. If we heat *two* grammes of water 1° C we shall need 4·2 joules (or 1 cal) to warm the first and another 4·2 joules (or 1 cal) to warm the second, which will make 8·4 joules (2 cal) altogether. If we had 50 g of water and warmed it 1° C, each gramme would need 4·2 joules and so we should use 50 × 4·2 joules (or 50 cal). If we now warm the 50 g of water another degree we shall use another 210 joules, making 420 joules (how many calories?) in all. So to work out the amount of heat energy you need to warm water we just multiply the number of g of water by the number of °C we warm it, and this gives the answer in calories. If we want the answer in joules we then multiply the number of calories by 4·2.

Questions

How many joules (or how many calories) of energy are needed to warm:

(1) 12 g of water up 10° C?
(2) 20 g of water up 13° C?
(3) 10 g of water from 5° C to 15° C?
(4) 40 g of water from 12° C to 62° C?

How much heat energy is given out when:

(5) 60 g of water cool 5° C?
(6) 8 g of water cool 15° C?
(7) 20 g of water cool from 100° C to 20° C?
(8) 50 g of water cool from 97° C to 15° C?
(9) A gallon of water weighs about 4500 g. How much heat energy is needed to heat it from 20° C to the boiling point?

You can see that if you have a lot of water to warm you will get some very big numbers in your answer. Scientists often give big answers in kilojoules (kJ) or kilocalories (kcal). How many joules will there be in a kilojoule? (Think about a kilometre.)

You probably know that when you have a bath you should always run some cold water into the bath first and then turn on the hot tap. (This is to avoid damaging the enamel by suddenly expanding it.) When you do turn on the hot tap the cold water warms up. Where does the heat energy that warms the cold water come from? Where does the heat from the hot tap water go? Now that we can measure heat energy we can find out something about this by doing experiments.

Experiment: *To find out where the heat energy goes when hot and cold water are mixed*

Before you start work you must do a very easy sum. The density of water is 1 g/cm³. How many cm³ of it must you measure out to get 100 g of it?

Use a measuring cylinder to measure out 100 g of water and pour it into the metal can provided. (What precautions should you take when you measure it?) Put the thermometer in the water and leave the can on the bench. Pour another 100 g of water into your beaker and start warming it over a smallish non-luminous flame.

Read the temperature of the water in your can and record it. Transfer the thermometer to the beaker and go on warming the beaker until its temperature is between 50° C and 60° C. Then turn the burner out. Stir the water very carefully with your thermometer. The temperature will probably go on rising for a little while. (Why? Where is the water getting this heat energy from?) When it seems about steady, read and record the temperature of the hot water. Using a handkerchief, if necessary, to hold the beaker, pour the hot water into the can of cold water. Stir very carefully with the thermometer while you count twenty fairly slowly. Then read the temperature again. Remember to read between the divisions.

After they are mixed, the can contains all the original hot water and all the cold, and both lots of water finish up at the same temperature. So you can work out how many degrees the hot water has cooled and how many the cold water has warmed up. The table of observations in your book should look like the one below.

Mass of hot water used = g
Mass of cold water used = g
Temp. of hot water = °C
Temp. of cold water = °C
Temp. of mixture = °C
The hot water cooled down °C
The cold water warmed up °C
The number of joules (or calories) lost by the hot water was .
The number of joules (or calories) gained by the cold water was .

You see that you have used the metal can to help you to measure numbers of joules gained and lost. A can or other vessel used in this way is often called a *calorimeter*, so if you meet the word you will know what it means.

When you have finished doing your sums try to answer the following questions:

(1) Add together the temperatures of the hot water and the cold water and divide by two, which gives a temperature half-way between them. Was the temperature of the mixture roughly half-way between those of the hot and cold water?

(2) Was the number of joules (or calories) lost by the hot water either the same as that gained by the cold water? or a bit more than this? or a bit less?

(3) Was anything else heated up besides the cold water? (If so it will have used up some of the hot water's heat.)

(4) Did anything else cool down as well as the hot water? (If so it will have helped to warm the cold water.)

(5) Did any of the hot water's heat, do you think, escape into the air in the lab.?

Think carefully about the answers to these questions and try to decide whether it is roughly true to say that:

If no heat is lost during the experiment, then

Heat lost by hot things = Heat gained by cold things.

The words 'if no heat is lost' are very important and so before we do any more experiments to measure heat energy we must find out something about how heat can be lost – in other words how it travels from one place to another. When we know more about how heat travels we shall perhaps be able to stop it from travelling away from our apparatus so that we shall lose less of it when we experiment.

Chapter 15 How Heat Travels

Experiment: *To try the effect of heating and cooling metal and glass suddenly*

Arrange a Bunsen burner to give a large roaring flame. Hold a piece of glass rod in a pair of tongs and put it quickly into the hottest part of the flame. (Where is this?) It is possible that the glass will crack. Repeat this using a nail. Does the nail crack? Take another piece of glass and hold it above the flame, turning it gradually to and fro so that it warms up slowly, and finally putting it into the hot part of the flame. Does the glass crack? If it does, try again more slowly. When the rod is red hot plunge it quickly into cold water (or put it under a running tap). Do not put your face near the glass when you do this. Does the glass crack? Repeat the experiment with the nail. Does the nail crack?

You may have known before you did the experiment that glass is likely to crack if you heat it or cool it too suddenly. When you wash up you don't put a thick glass bowl into very hot water (unless it is made of special glass). We are now going to do experiments to find out why some substances crack more easily than others.

Experiment: *To compare how heat travels through iron and glass*

Arrange your Bunsen burner to give a small roaring flame (how?) and put the asbestos square on the left side of it to protect the bench. Take the nail in your left hand and the piece of glass in your right hand and hold each by one end. Move the other ends into the flame slowly so that the glass doesn't crack, and adjust them until they are just at the top of the blue cone. (Why?) Keep hold of the glass and nail until one of them is too hot to hold, then drop both on the asbestos square. If you miss, pick up the hot object quickly with tongs and put it on the square.

Fig. 15.1. This boy is comparing how glass and iron conduct heat. Which hand will let go first?

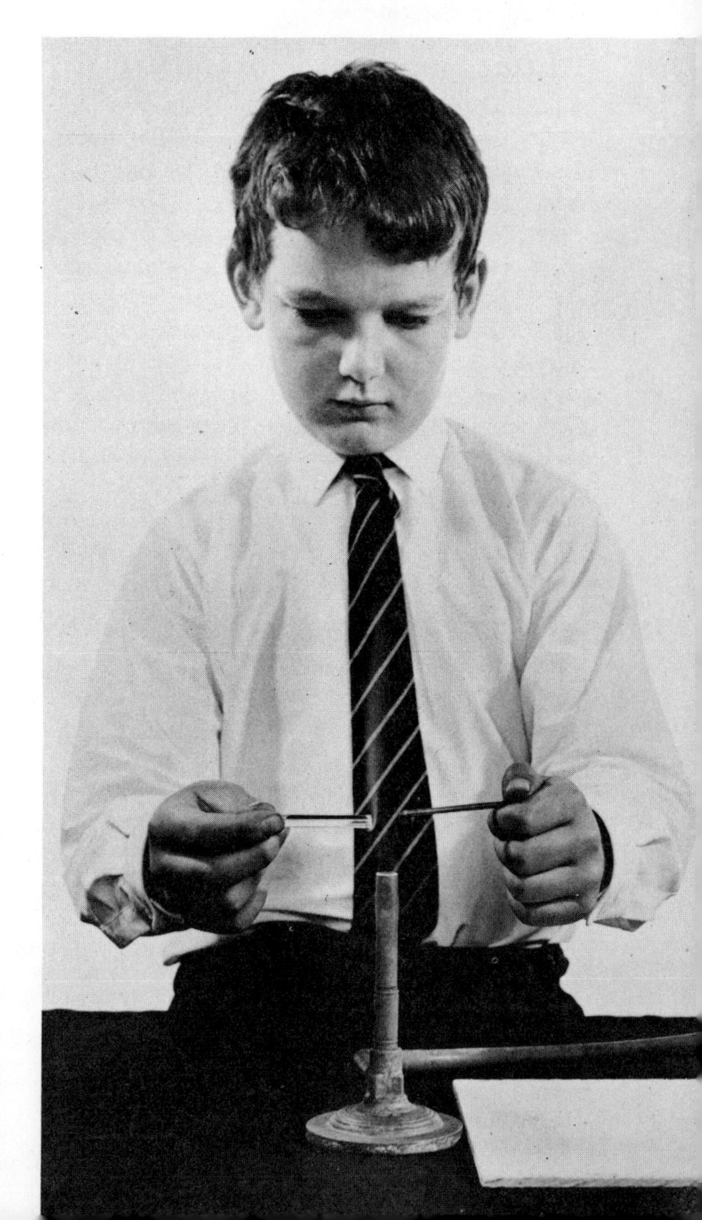

Does the heat travel more quickly through the glass or through the nail? Did the glass get red hot? Did the nail?

This simple experiment shows that heat energy can travel along a substance and that it travels better through some substances than others. When heat travels through a substance in this way we say that it is *conducted*, and we call the process *conduction*. Substances that allow heat to be easily conducted through them are called good conductors: if they do not we call them bad conductors or sometimes *insulators*. Most metals are good conductors, while substances like glass, brick and wood are bad conductors, and others like felt or cotton wool are insulators.

More about molecules

When a substance is heated we believe that the molecules in it move about faster. So if we heat one end of a substance strongly its molecules move about violently. If they do this they will set nearby molecules moving too. If you are waiting in an orderly queue to go into a room or a cinema and some people near you begin to struggle or push or move violently about, then their movement soon affects those near them and may disturb the whole queue. Sensible people would of course try to move away from the trouble, but molecules can't think and so the disturbance passes along the substance. Each molecule passes on some of its energy to the next and so the substance gradually becomes hot all through.

We can now try to explain why it was that in our first experiment the glass cracked but the nail did not. Can the difference be because iron conducts heat well while glass is a poor conductor?

Fig. 15.2.

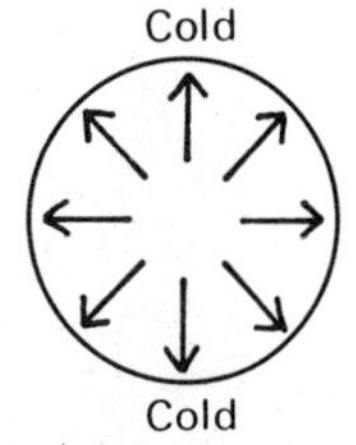

Can heat get out quickly enough?

Suppose Fig. 15.2 represents a hot glass rod that is suddenly cooled at its surface. Before the whole rod can cool, heat must travel from the inside to the outside. Can it do this easily? If not it will stay hot inside while the outside cools. What happens to a hot solid when it cools? If you try to stop a hot body from contracting, what happens? (Think of the force that was needed to break the cast-iron rods in the experiment on p. 33.) Do you think that this is the sort of force that cracks the glass? What do you suppose happens when you heat the glass suddenly?

Why didn't the iron crack – it was just as hot? Can heat escape from the inside of the iron more easily than from the glass? Why?

Is a thick piece of glass less likely to crack than a thin one? Is the glass more likely to crack if you cool it more? Why? If you had a very big block of iron and heated it up very much, do you think it would crack if you cooled it suddenly?

If a substance cracks easily when you cool it or heat it suddenly it is because (1) it contracts when cooled and expands when heated, *and* (2) it is a bad conductor of heat. Metals do not easily crack because they are good conductors. Special oven glass dishes do not crack easily because they hardly expand at all and so not much force is produced when the temperature changes. There is a substance called silica that expands so little that it can be heated red hot and put straight under a tap without cracking.

Questions

(1) If you want a house to be warm in winter, would it be a good idea to make the walls of metal?

(2) Next time you visit a real old country cottage or an old castle, look at the walls. They are usually much thicker than the walls of a modern house. Do you think that such walls help to keep a building warm in winter? Why?

Fig. 15.3. Porchester Castle from the air. The castle walls were built very thick to prevent enemies from battering them down. What other advantages does a thick wall give – in summer? in winter?

(3) You may have thought that it would have been good fun to live in the days when people wore suits of armour. Do you now think that it would have been so pleasant? What would it be like inside a suit of armour in winter? in summer?

Experiment: *Another way by which heat travels*

Fig. 15.4.

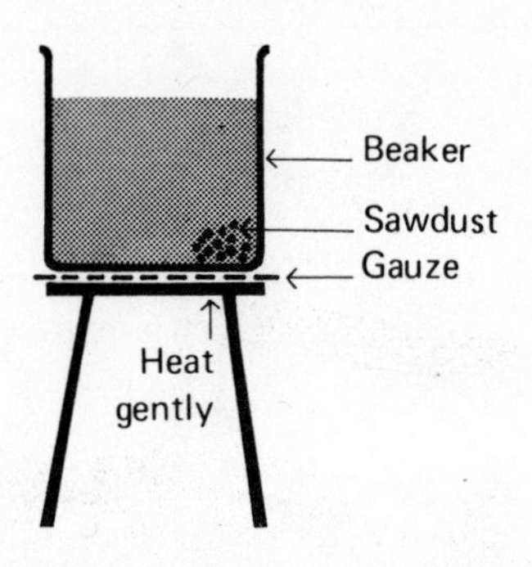

Arrange an unlit Bunsen burner underneath a tripod and gauze and then place on top a beaker about two-thirds filled with cold water. Leave the beaker until the water is still, then take a pinch of wet sawdust between your fingers and drop it *gently* down one side of the beaker, disturbing the water as little as you can. Most of the sawdust will probably settle to the bottom of the beaker. Remove the Bunsen, light it and adjust it to give a small non-luminous flame. (How?) When the water is still, again put the burner below the beaker so that it is just below the sawdust. Watch the beaker carefully to see what happens. You will probably find that the outside of the beaker gets misty and you can clear enough surface to see through with blotting paper or filter paper, but try not to disturb the beaker.

You will see currents of water moving about in the beaker. Copy Fig. 15.4 into your note-book and mark on it how the currents travel, using arrows to show how the water is moving.

What do you think would have happened if you had heated the side of the beaker opposite to the sawdust? If you have time, try this.

What happens to a liquid when you warm it? If the volume of 1 g of water gets bigger when it is warmed, does the water get more or less dense? Does this explain why the warm water rises? Would you expect any currents in the water if you heated the *top* of the water instead of the bottom? (You can try this experiment but you will have to fill the beaker to overflowing or you will probably crack the glass at the top with your Bunsen flame.)

The currents that we see in water or in other liquids when they are heated at the bottom in this way are called *convection* currents, and this is another way in which heat can travel from one place to another. Another way to show the currents is to drop carefully into the water a small crystal of potassium permanganate. You can then follow the movement of the water from the colour.

Something to do at home

Problem: **Is it possible to make a convection current by *cooling* water? Take home a few crystals of potassium permanganate and you may be able to solve this. Take a jug or a glass bowl or pie-dish and pour in water until it is about 2 cm from the top. While the water is getting still, think out what to do. An ice cube from the fridge is convenient for cooling the water. You will drop into the still water at one side, very carefully, one crystal of permanganate. Whereabouts should you cool the water if it is to circulate? – at the top or at the bottom? Put the ice cube in the water gently in the most suitable place and see what happens.**

Experiment: *To show that convection currents are formed in air when it is heated*

Light a Bunsen burner and adjust it to give a small non-luminous flame. Light a small piece of blotting paper and as soon as it has caught, blow it out. Hold the smouldering paper a little way above the burner but not near enough to the flame to catch fire. Watch what happens to the smoke from it. The smoke goes where the air from the burner takes it. Do you see a convection current? If there are many people doing the experiment you may not be able to see whether any of the smoke comes down again.

Something to do at home

A model

Fig. 15.5.

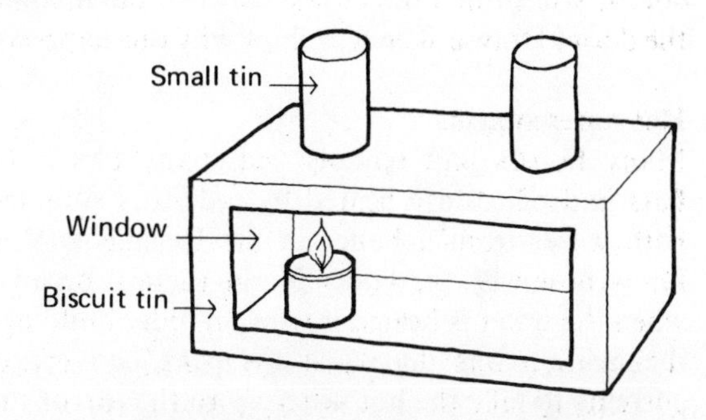

Get an old tin with a lid and make two holes about 2 cm diameter in the lid with a fair gap between them. Remove the tops and bottoms from two other small tins and stand them on top of the two holes as shown.

Put a small piece of candle in the large tin below one of the two holes and light it. Replace the lid. Stand the other two tins in position. Light a piece of old rag or blotting paper, blow it out and hold it over the small tin that is *not* above the candle.

What happens to the smoke at first? Watch carefully the tin above the candle. After a few seconds, what do you see coming out from the top of it? Does the experiment show that the burning candle starts a convection current in the air?

If you are good at metal work (or if big brother or father is) you can solder the two small tins in position and you can cut a window in the side of the big tin and cover it with polythene held in place with Sellotape. This will let you see what goes on inside the tin. Or you can use a wooden box with a glass window instead of the large tin.

In liquids and in gases the molecules are able to move freely about and we see from these experiments that when we heat a gas or a liquid the molecules do not just pass heat from one to another – they move themselves and take their energy with them. This movement we call a convection current.

Questions

(1) If you have a fridge at home, look inside it for the freezer, which will be somewhere near the top. Why is it put at the top? What will happen to the air that is cooled by the freezer? What would happen to the food at the top if the freezer was at the bottom?

(2) If you have a coal fire at home you will have noticed that smoke and flame go up the chimney. This makes a convection current. Some cold air must come in somewhere. How do you think it gets into the room? How will it get across the room to the fireplace? Why do your feet sometimes feel cold if you sit in a direct line between the fire and the door? Draw a sketch to show why this happens.

Hot water systems

Many houses and schools and many blocks of flats and offices are heated by radiators supplied with water from a boiler in the basement. You know now why the boiler is put there – because when the water is heated it rises. In older buildings the heating was done just by using convection currents to take the hot water up to the top of the building, thus heating the rooms on its way. Nowadays it is usual to help the water by pumping it as well but it is still best to put the heating plant at the bottom so that the heating helps the pump.

Fig. 15.6. A modern refrigerator. When food goes bad it is because chemical changes take place in it. By keeping the food cold the changes are slowed down or even stopped completely, so the food keeps longer. Why is the freezer unit put at the top of the fridge?

Fig. 15.7.

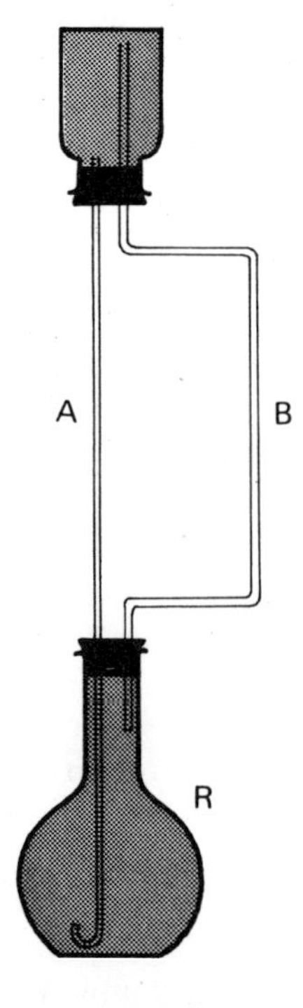

Figure 15.7 is a model to show how such an arrangement works. R is a cold water tank, A represents the pipe through which water enters the bottom of the boiler and B the pipes through which

the hot water rises. Notice that the cold pipe runs down to the bottom of the flask and that the other pipe starts from the top of it. If a little potassium permanganate is poured into R and the flask is heated you can follow the path of the water round the system.

Do liquids conduct heat?

Our experiments show us that heat will travel through water quite well by the method of convection. Does water also conduct heat? We can perform a simple experiment to find out something about this.

Experiment: *To find whether water conducts heat well or badly*

Fig. 15.8.

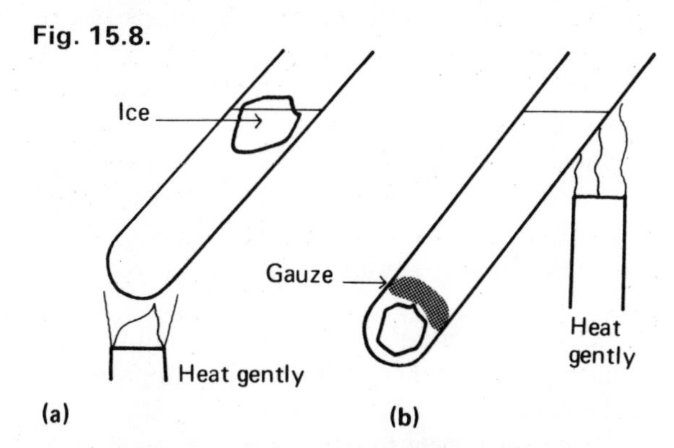

Take two boiling tubes and in one put water to about 5 cm from the top. Put a small piece of ice on top of the water. Hold the tube in a pair of crucible tongs and warm the bottom of it with a Bunsen burner (Fig. 15.8 (*a*)). Watch the ice and see whether it melts quickly – you can time it if you like. By what methods does heat reach the ice when the tube is heated at the bottom?

If we want to stop convection currents from being formed in the water we must heat the water at the top and put the ice at the bottom. Ice floats, so we use a small piece of bent wire gauze to keep it down. Arrange your apparatus as in Fig. 15.8 (*b*), and heat the tube at the top. Watch the ice. Does it melt quickly? Can you boil the water at the top of the tube while the ice remains at the bottom? Can heat reach the bottom of the tube by convection? Does water conduct heat well or badly?

Most liquids (other than liquid metals) are bad conductors of heat. Gases are even worse. This is not surprising, because the molecules in gases and liquids are usually further apart than in solids. It is thus more difficult to pass on energy from one to the next, so they don't conduct well. On the other hand, since the molecules can move about freely it is easy for molecules with much energy to move to another place. So both gases and liquids allow heat to pass through them quite easily by convection.

Keeping warm

You know that people wear clothes in temperate climates like Britain's mainly to keep warm. Inside us chemical changes happen to the food we eat and provide us with energy to do things and heat energy to keep us warm. If we do not get enough food to eat we cannot keep warm enough to live. Clothes help to stop this energy from leaking away from us and so we wear thicker clothes and different materials in cold weather.

Experiment: *To compare the way in which different materials stop heat from travelling*

Fig. 15.9.

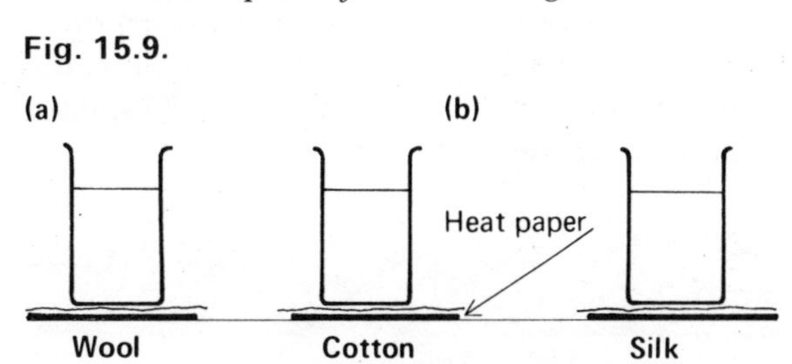

To do this experiment we use heat paper. This is paper that has been soaked in a chemical substance called cobalt chloride. (How many elements are there in this substance and what are they?) This substance is pink when cold but turns bluish-green when hot. Place some sheets of heat paper on the bench and on each sheet put one of your samples of material. On each piece of material stand a calorimeter. Heat some water in a beaker or a kettle and when really hot pour some quickly into each calorimeter so that they are about half full. Leave the calorimeters for two minutes and then lift them off the materials. Look at the heat paper underneath.

Fig. 15.10.

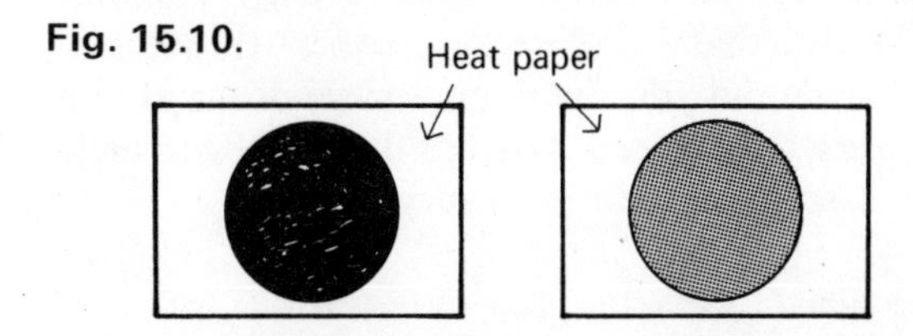

Are some of the sheets more green than others? What does this show? Which materials allow heat to pass through them easily? Were there any materials

under which the paper hadn't changed colour at all? Is the change in colour a physical change or a chemical one?

Make a list in your book of the materials that transmit heat well and those that transmit badly. Try to arrange them in order of merit. You can test the good insulators again and try to get an order by leaving the hot water over them for a longer time.

You can try a similar experiment with solids like plastic, glass, metal and so on. If the test is to be really fair you must choose samples that all have the same . . . what? Scientists use the results of experiments like these to decide what substances are good for making walls for houses or lagging for hot water tanks.

Animals keep warm without clothes but those that live in cold countries grow fur or hair, and birds grow feathers to keep them warm. These coverings are poor conductors of heat and so the heat of the body is kept in. Men and women often wear clothes made of sheepskin or fur because they keep them warm. Why are these substances such poor conductors?

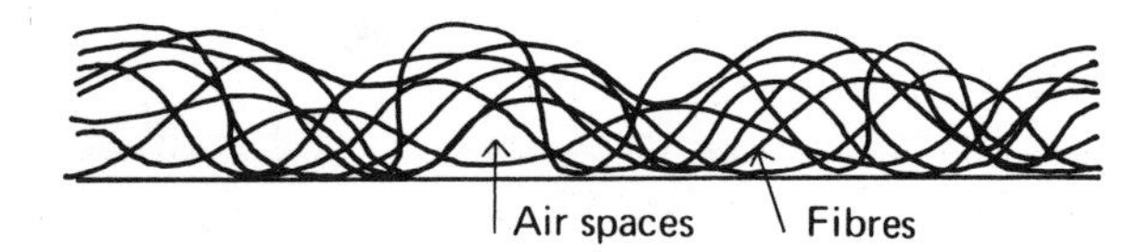

Fig. 15.11. A diagram to show what the wool on a sheep's back looks like.

The wool on a sheep's back consists of a great many hairs that entangle many air spaces among them. Because the hairs are there it is difficult for convection currents to start so the sheep does not lose heat easily that way. Because air is a very bad conductor of heat little heat is lost by conduction so the animal keeps warm.

If in cold weather you put your hand on a piece of woollen material like flannel or on a sheepskin coat it feels warm to the touch because it does not allow your hand to lose heat energy through it. But if you put your hand on a piece of metal you feel very cold indeed, because the metal conducts heat away from your hand very rapidly.

Questions

(1) Suppose that you are at an open-air swimming pool on a very hot day when the sun is shining on the stonework. What difference do you notice between putting your bare foot on the stone and putting it on a towel lying on the stone? Which feels hotter to your foot? Why?

(2) A metal coffee pot is often fitted with a wooden or plastic handle. Why?

(3) Should the following be made of substances that transmit heat well or badly: blankets, house-walls, the tops of cooking stoves, the walls of a fridge, saucepans, kettles. Give reasons for your answers.

(4) In very cold weather birds are sometimes seen to fluff up their feathers. Why do you suppose they do this?

(5) String vests seem to have very little material in them. Why do they help to keep you warm?

(6) Is it warmer to wear two thin pullovers, one over the other, or one thick one? Explain.

(7) On a winter evening the room usually feels a good deal warmer when the curtains are drawn. Give three possible reasons why this is so.

Fig. 15.12. Keeping the heat where you want it. A great deal of energy is needed to heat a house so it is sensible to try to keep the heat in. Why is fibreglass such a good insulator? Is the house shown a modern one or rather an old one?

Another way in which heat travels

When you stand in the sunshine you feel the warmth of the sun, but if you move into the shade of a building you feel much cooler. When you stand in the sunshine energy certainly reaches you from the sun. You know, I expect, that for most of the way between the sun and the earth there is just empty space – about 140 000 000 kilometres of it. You have, I am sure, seen pictures of space ships and of men outside them working in space. Since space is empty there are no molecules in it so there can be no convection currents and no conduction either. There must be another way by which energy can travel and we call this way *radiation*. When this radiated energy falls on our bodies it warms them up. In winter especially, the air can be quite cold and it may be freezing in the shade, yet when you stand in the sunshine the radiation makes you feel quite warm. We will now try to find out more about this mysterious way by which energy can travel through empty space. We know that radiation can travel through air as well as empty space because we feel the sun's warmth as we walk about in the air, and of course we shall have to do our experiments in the air.

Something to do at home

To show that the sun's rays carry a great deal of energy

Warning: **Never look directly at the sun and never look at it through a lens. You may seriously damage your eyes if you do this.**

Fig. 15.13.

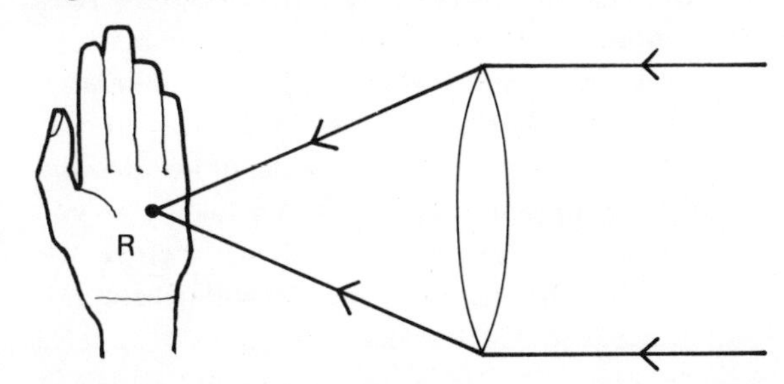

You will need a lens for this experiment. Hold your hand in the sunshine. It will feel warm. Take the lens and hold it between the sun and your hand. It will probably make a bright ring of light on your hand. Move it nearer (or further) from your hand so as to make the ring smaller. Your hand will probably at once feel warmer. Try moving the lens until the ring of light is really very small. Can you keep your hand there for long?

Suppose that R (Fig. 15.13) is the little bright ring on your hand. Why is the spot so much warmer than when the lens isn't there? If the lens was a wider one would you expect the spot to be hotter or colder? (What name do scientists give to the width of a lens?)

Lenses have been used in this way for many centuries and are known as burning glasses because they can be used to light a fire (if the sun is shining!) You should never leave a lens where the sun can shine through it onto a table or onto inflammable material such as paper. Fires have been caused by this happening when no one was around. Your experiment shows how much radiated energy from the sun warms things when it falls on them. We say that the things *absorb* the radiation.

Fig. 15.14. Using the sun's energy. The sun sends much energy to us by radiation. In desert countries there is no wood for fuel and some people use cookers like this one to boil water and to make stew or soup. How many mirrors are there? Why don't they use one big one? Why is the frame made so that it can turn round? Why does the man wear a white headdress?

Experiment: *To see if black and polished tins absorb radiation equally well*

Fig. 15.15.

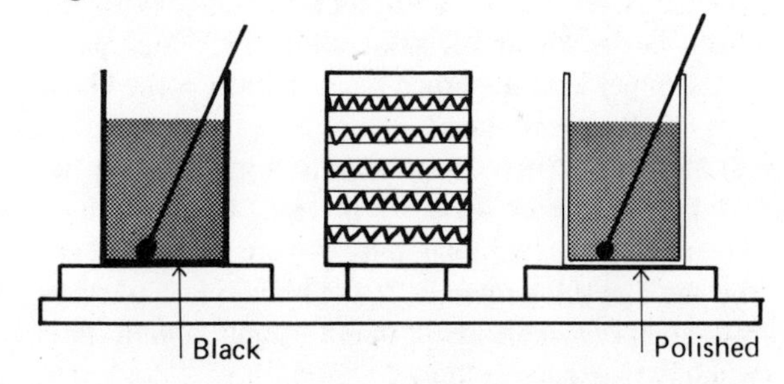

You will need two tins (or copper calorimeters) that are alike. Blacken one by holding it in the smoke from a burning candle. Polish the other as much as you can. Measure out 100 g of water (how many cm³?) into each, put in a thermometer and record the temperatures. Place the two tins at the same distance from the electric fire, switch on and leave them for 5–10 min. Read the temperatures again. Make a table for your observations as follows:

	Black	Polished
First temp.		
Second temp.		
Rise of temp.		

There are three ways in which heat might reach the tin. What are they? Is much heat conducted

through the air? Would convection currents help to heat the tins? Draw a sketch to show how the air will move. What will it do to the tin? Most of the heat energy reaches the tins as radiation. Are the black and polished surfaces equally good at absorbing radiation? If not, which absorbs best? What happens to the energy that isn't absorbed? Why did you put the same mass of water into each tin? Why did you put them at the same distance from the fire?

Something to do at home

Fig. 15.16.

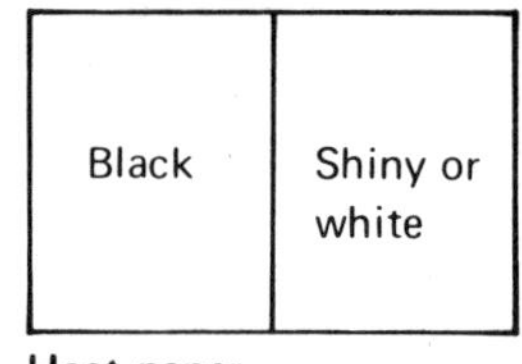

(1) Ask at school for a piece of heat paper. Paint half of one side of it black and the other half with some white paint – or bright aluminium paint, as in Fig. 15.16. Or you can paint a pattern, say your initials, on a black background. When the paint is quite dry hold the painted side of the sheet in front of an electric, coal or gas fire – not near enough to char it. After a short while look at the back of the sheet. Is the colour the same all over? Which parts of the paper have absorbed most radiation, the black parts or the aluminium?
(2) Next time there is snow on the ground and the sun is shining too, spread a piece of black cloth or paper on the snow in one part of your garden where the sun can shine upon it. If the black cloth absorbs more energy than the ground round about, how would you expect to know?
(3) If you have at home a reading lamp with an ordinary electric light bulb in it, you can try a simple experiment about radiation next time you switch it on – but it must be quite cold when you start. Hold your hand two or three cm from the bulb. Switch on, count three and then switch off again. Did your hand feel warmed when you switched on? Now touch the bulb. Does the bulb feel warm? In such a short time could the bulb have warmed the air round it by conduction or convection if it wasn't warm itself?

The energy that reached your hand must have travelled through the glass of the bulb. You know that glass is a very poor conductor of heat, so the heat didn't reach you by conduction – and anyway the glass wasn't warmed, so it couldn't have passed heat to the air. So it was *radiation* that you felt and you have learned that radiation can travel through glass and through air.

Fig. 15.17.

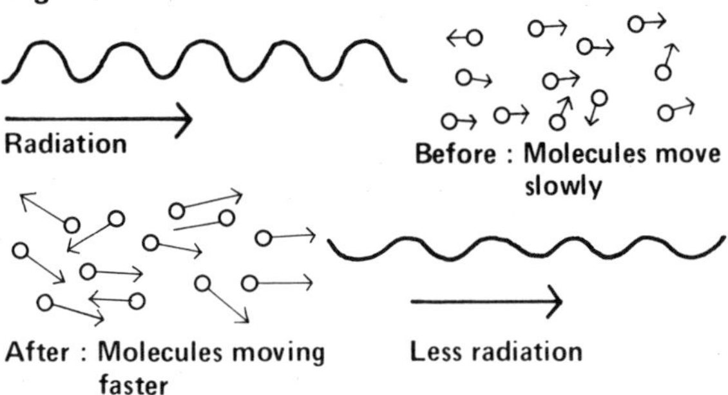

Scientists believe that radiation is made up of waves, like the radio waves that carry our TV programmes only very, very much shorter. They are so short that they can find their way in between the molecules of substances that are in their way. However, the molecules *do* get in the way and so some of the energy is stopped by the molecules and warms them up – makes them move faster. So you feel radiation from the electric bulb before the glass gets hot, but in time the bulb will get too hot to hold because of the energy it stops.

More about radiation

We have discovered that a black surface absorbs radiation better than a polished one. It would be interesting to find out whether black and polished surfaces are equally good at sending out radiation.

Experiment: *To compare the rates at which black and polished surfaces give out radiation*

Figure 15.18 shows how this can be done. The tin has one half of the outside polished as much as possible and the other half is painted dull black. It stands on cotton wool, or a cork, inside the beaker, and the two halves of the beaker are separated from each other by cotton wool, as shown. The tin has a bung and can be covered over with cotton wool. Two thermometers are arranged in the two spaces but do not touch the sides of the tin or the beaker. Check that they read the same before you begin and record the temperatures. Pour hot water into the tin, put the bung in and cover the top with cotton wool. Wait for a few minutes and then read the thermometers again and record the readings.

Do the thermometers still read the same? If not, which one has received most energy? How has the energy reached it? Do the two surfaces radiate energy equally well? Why do we stand the tin on cotton wool?

Fig. 15.18.

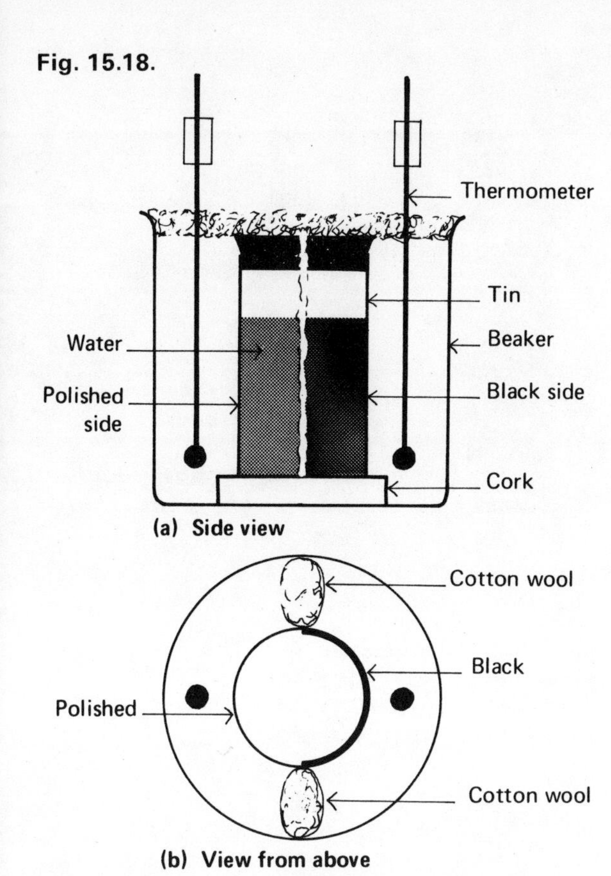

Fig. 15.19. A Crookes radiometer.

Why is cotton wool put down the sides and over the top of the tin? What might happen if we left the wool off the top or didn't use a bung?

Something to do at home

Take two similar tins and polish one of them as well as you can. Paint the other dull black. Make marks in each tin at the same distance above the bottom. Stand each tin on a thick pad of folded newspaper (to protect what is beneath) on a table or bench. Pour hot water from a kettle or jug into each tin up to the mark. Leave them for a few minutes and then read the temperature of the water in each. Which of them has cooled most rapidly? What does this show?

A toy that works by radiation

Figure 15.19 shows a little toy that works when radiation from the sun or from a fire or lamp falls on it. It is called a Crookes radiometer. Inside the glass bulb there are very few air molecules. The little paddle wheel has one side of each paddle polished and the other black. When the sun shines on it it turns round quite fast.

If it is turning fast, put a cardboard box over the top of it to shut out the sunlight and leave it for a few minutes. Take the box away and see whether it starts to turn with the black side first or the polished side first.

What happens to a body when radiation falls on it? Will the black side or the polished side of each paddle get warmed up most? The air molecules hit both sides of the paddle. Will they bounce off faster from the hotter surface or the cooler surface? Which side of the paddle gets the harder bangs from the air molecules? Which way would you expect the paddle to go round? Does it?

The little paddles are usually made of a substance that is a bad conductor of heat. Why wouldn't the toy work so well if a very good conductor were used?

Keeping things hot (or cold)

We have already talked about how we keep ourselves warm by wearing clothes that stop heat energy from leaving us by conduction and convection. There are other things that we want to keep hot besides our bodies, and we want to be able to keep things like ice-cream cold. If we want to stop heat from moving, we must do our best to prevent convection, conduction and radiation. You may know that a vacuum flask does this very well. This is how it works. (Fig. 15.20.)

The metal case of the vacuum flask is put there to protect the glass part inside, because glass is easily broken. The flask is made by putting one flask inside another, thus providing a double wall. Then the inside surfaces are silvered by using chemicals. Finally the air is sucked out from between the two walls and the space between them is sealed off. You can usually find somewhere at the bottom of the flask the little spike S showing where this was done.

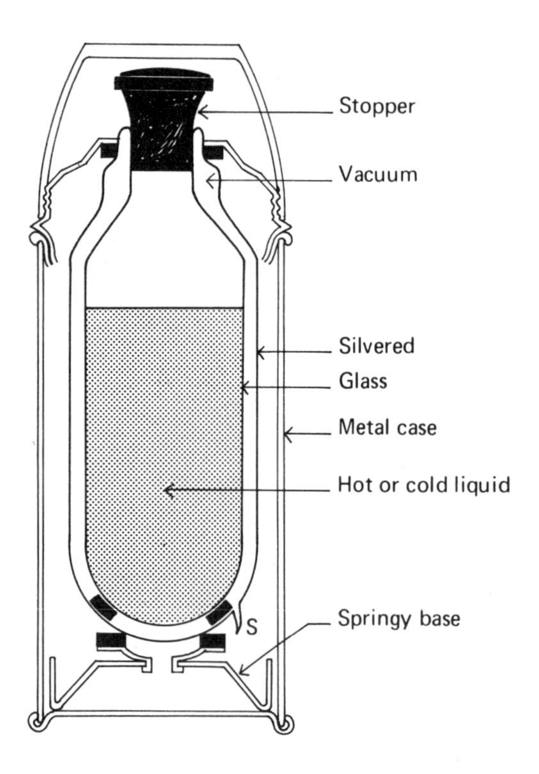

Fig. 15.20. A diagram of a vacuum flask.

Fig. 15.21. An electric fire. The fire changes electrical energy into heat energy. By which ways does heat energy leave the bar to warm the room?

It is now very difficult for heat energy to get into or out of the flask. A vacuum does not conduct, so heat can't get through the sides: silvered surfaces do not radiate much, and provided you put the cork in there can be very little convection. Almost the only way heat can get in or out is by conduction up the glass neck of the bottle or through the cork – and both glass and cork are bad conductors. The flask was first invented to keep things cold, and it does this just a little better than it keeps them hot. Why? (*Hint:* think how heat would travel in and out if you left the cork out.)

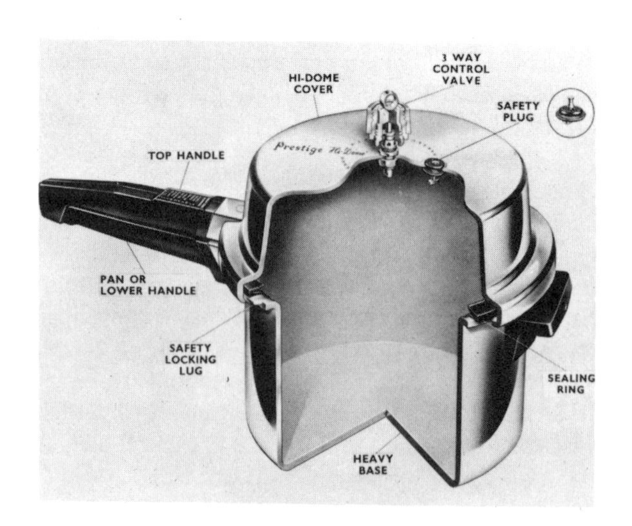

Fig. 15.22. A pressure cooker. When water boils at high pressure it boils at a temperature above 100° C, so food cooks more quickly in a pressure cooker. Why is the handle not made of aluminium like the rest of the cooker? What is the safety plug for? How does it work?

Questions

(1) Why is a tea cosy often put over the teapot? How does it work? How would the pot lose heat if you left it off?
(2) Look at the hot water tank in your home and make a sketch to show how the heat is kept in.
(3) In tropical countries people generally wear white or light coloured clothes. Why is this?
(4) Houses in hot countries are frequently colour washed with white or light colours. Why?
(5) How does heat reach a saucepan when it is put on an electric cooker? How does the pan lose heat? Why is it good to keep your pans shiny on the outside?
(6) Space ships and satellites make use of radiation from the sun to keep their batteries charged. What would you think would be the best sort of surface to expose to the sun for this purpose?

Fig. 15.23.

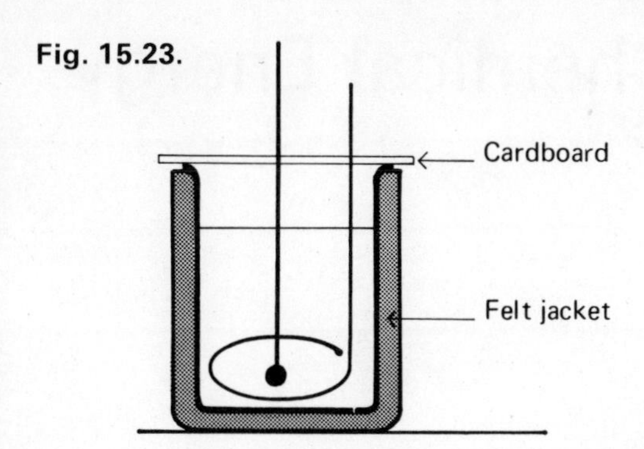

(7) Figure 15.23 shows a calorimeter arranged for doing an experiment in heat measurement. Why has it a felt jacket? What does the cardboard on top do? Do you think that the calorimeter should be kept highly polished or blackened? Why? Why is the piece of bent wire there?

(8) Find out who invented the vacuum flask and what he wanted to keep in it.

Summary

There are three ways in which heat energy travels from one place to another:

Conduction: Each molecule passes on some energy to its neighbours.

Convection: The molecules move and take their energy with them.

Radiation: The energy travels through the spaces between the molecules and the molecules collect some of it as it goes through. Radiation is like radio waves only very much shorter and this is why it can get between the molecules.

Chapter 16 Heat and Chemical Energy

We know that when some chemicals react together they give new chemicals and heat energy. Give two chemicals which do this. We say that

$$\text{chemical energy} \xrightarrow{\text{gives}} \text{heat energy.}$$

We must now go into this more deeply.

Heat energy is so useful – think of the ways we use it to cook our food and to keep us warm. How do you cook at home and heat your house? You may do this by burning something. The chemicals you burn to give heat are called *fuels*. Perhaps you use electricity. Electrical energy is made at a power station and fuel is used there. How does your fuel get to you? By tanker or lorry or through pipes under the ground? We will now try to find out how much heat our fuel is able to give.

From gas

Light a Bunsen burner to get a small non-luminous flame and put it under a tripod and gauze. Pour 100 cm^3 of water into a beaker and then take its temperature. Put the beaker of water on the tripod and then read the time to the nearest second. Gently stir the water and after two minutes take its temperature again. Now you can work out how much heat the gas gives in two minutes.

Mass of water = g
Rise in temperature = °C
Number of joules =

You can read how many joules are needed to heat 1 g of water 1° C on p. 111.

Did all the heat from the gas go into the water? How can you change the apparatus to get a better answer? Would a bigger beaker be better? Would it be better to use more water? If you did the same experiment using a bigger flame would you get the same answer?

In the next experiment we are going to find how much heat is given by 1 g of fuel and this means that we have to weigh it carefully.

More about weighing things

The picture on p. 125 shows a *top pan balance*. It looks different from the balance you have used so far but the rules for using it are just the same. Look them up on p. 13. Ask your teacher how many ice creams you can buy for the price of one of these balances; they cost a lot of money and you must be very careful with them. They can measure very tiny weights – just like a microbalance. What do we call a balance which can detect very small changes? Does the balance in your laboratory show weights to 0·001 g (1 milligramme) or to 0·01 g (10 milligrammes)? These balances are so sensitive that they weigh differently if you walk near them. So here is another rule: When you are waiting to weigh, move about as little as possible.

When you have been told how to use the balance, practise by weighing the coins from your pocket.

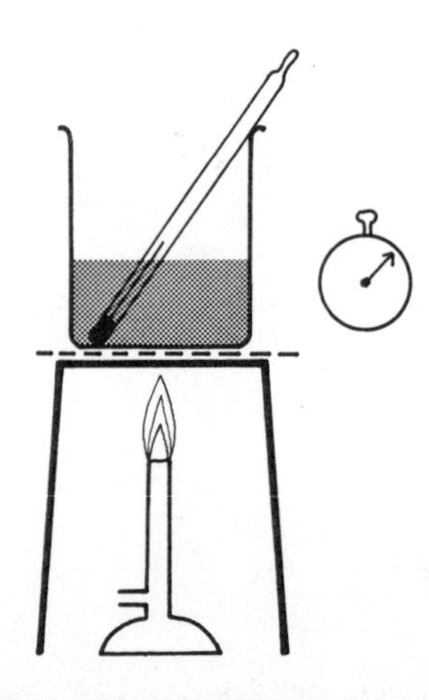

Fig. 16.1. Measuring the heat from gas.

Write down the results like this:

Mass of the coin is 6·233 g

Of course you will not get the same number. Now give the coin to your friend and ask what he gets for it (be sure to get it back!). It won't be long before you are very quick at using the new instrument.

Heat from meths

On p. 26 you will see a picture of a spirit lamp and you can use one like this to find out the heat coming from 1 g of meths.

First of all make sure that there is some meths inside the lamp and check that the wick works by lighting it for a few seconds. Now put the flame out and weigh the lamp on the top pan. Pour 100 cm³ of water into a tin and measure its temperature. Put the tin on a tripod (no gauze) and put the lamp under it so that the wick is about 3 cm from the bottom of the tin (see Fig. 16.3). Light the lamp and stir the water gently. When the water is about 50° C take a careful reading of its temperature and then blow the flame out. Weigh the spirit lamp again. You can now calculate how much heat 1 g of meths gives.

Fig. 16.2. Here is a top pan balance. Where do you put the thing you are about to weigh? Can you see the feet of the balance? Why can we raise or lower each one? It is most important to use a balance carefully and to keep it tidy and clean. What would happen if we put dirt on the pan? or scratched it?

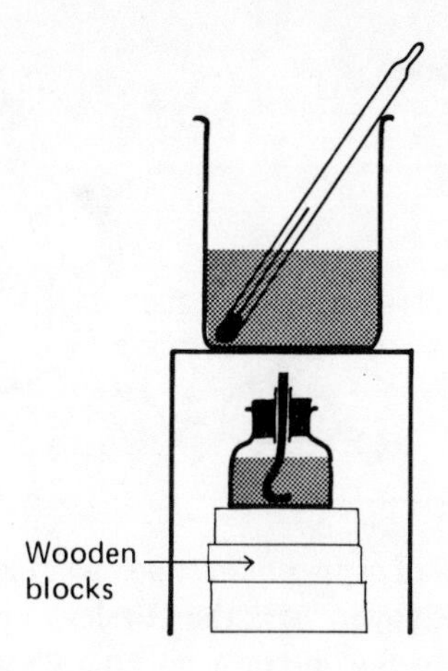

Fig. 16.3. The heat from meths.

Mass of the burner at the start	=	**g**
Mass afterwards	=	**g**
Mass of meths burned	=	**g**
Temperature of the water		
before	=	**°C**
after	=	**°C**
The rise in temperature was	=	
Number of joules taken in was	=	
Number given by 1 g of meths	=	

Would you get a bigger or a smaller answer if you used a shiny tin? Would it help to wrap paper round the sides of the tin? Can you suggest other ways of getting a better answer?

What happens when iron filings react with copper sulphate solution? If you have forgotten do the reaction in a test tube. Think how you would find out how much heat is given when 1 g of iron reacts in this way. Do you have to weigh exactly 1 g of iron? When you calculate the joules you may pretend that the copper sulphate is as easy to heat as water, and that the copper made by the reaction does not take any heat (after all it won't take much). When you have a way of doing the experiment you may be allowed to do it – if you have found a good way.

Fast and slow reactions

With meths and gas the reactions go at a good speed so that we can use the heat they give. But some reactions are not like this. Some go so quickly that it is hard to use the heat and these reactions can be dangerous. Then there are others which are so slow that you may not even notice the heat they give.

A fast reaction

Fig. 16.4.

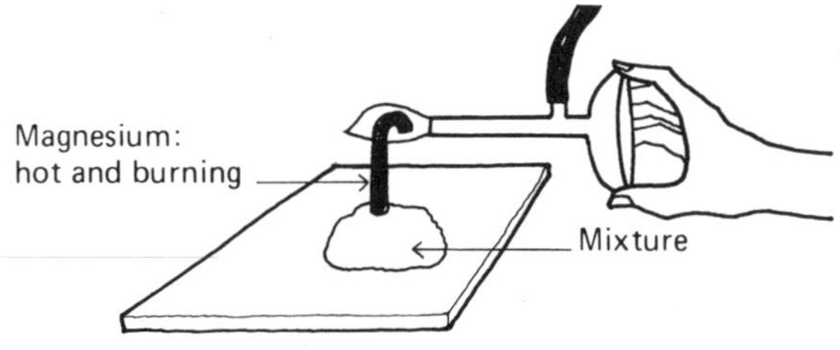

Put a spatula of copper oxide and one of zinc powder on a piece of paper. Mix the powders well with the spatula. Heap the mixture on an asbestos square and put a piece of magnesium ribbon (5 cm) in to act as a fuse. Now light the magnesium with a Bunsen burner holding it at arm's length. Take the Bunsen away as soon as the magnesium catches fire.

A slow reaction

Fig. 16.5.

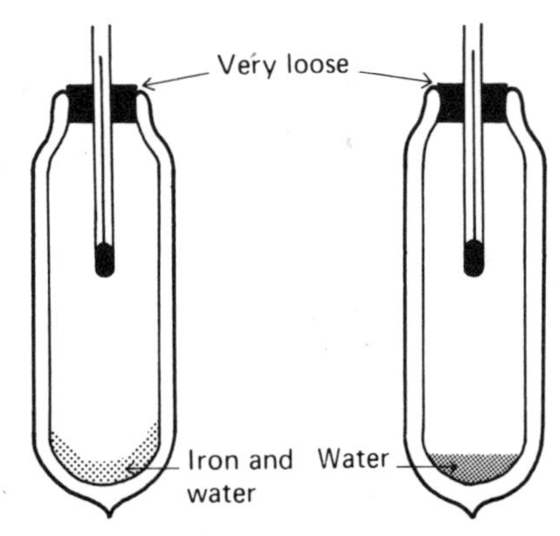

Put three spatulas of iron filings into a thermos flask and add 5 cm^3 of water (quarter of a test tube will do). Loosely cork the flask with a cork which has a thermometer set through it. Set up another flask without the iron filings but with the same amount of water. Place both flasks on the same window sill and take the temperatures every hour (see Fig. 16.5).

What was the thermos flask for? What would happen to the temperature if you used an ordinary bottle? Why did we use a second flask with only water inside it?

When you light a firework the chemicals inside it are changed and it gives out heat very quickly. A big Roman Candle makes enough heat to raise 100 g of water to the boiling point – and all in a few seconds. How many joules does it give if the water starts at 15° C? How silly it is to put fireworks in your pocket!

Fig. 16.6. Firework display. Energy in chemicals becomes changed very quickly into heat, light and sound energies in a most spectacular way.

If you live in the country you will know that it is very silly to stack damp hay. Very slowly, the damp hay takes part in a chemical reaction so that the middle of the stack gets warmer and warmer. Sometimes it even bursts into flame. Why is it that the heat cannot escape from the middle of the stack? (Fig. 16.7.)

Do you live where there are coal mines? All the waste is piled up into huge heaps called 'tips'. It is never safe to walk on these because slow reactions happen inside them, and they can get very hot indeed.

Scientists use a special word for reactions which give out heat – they call them *exothermic* reactions.

Burning – a special kind of reaction

To make heat in our homes we *burn* things. What happens when we burn magnesium ribbon? Is the change physical or chemical? When we burn coke or coal, gas or oil or wood or paper, the same kind of change happens. Your father would be delighted if he could get the fuel back again but he cannot! Burning is a chemical change.

But what happens during the change and what new things are made? What are flames made of and what is happening inside them as they give out heat and light energy? You know a lot about these changes already. For example, magnesium gives a white ash and paper a black one. But to learn more we will see how burning changes the *mass* of things.

What happens to the mass of magnesium when it burns?

Guess whether the white ash weighs more or less than the magnesium and then see whether you are right.

Tear 15 cm of magnesium ribbon into pieces about $\frac{1}{2}$ cm long and put them into a crucible (see Fig. 16.8). Put the lid on and weigh the whole thing on a top pan balance. Record the weight. Put the crucible on a triangle and tripod. Make your Bunsen give a small roaring flame and then put it under the crucible. (Be careful about strike back, see p. 26.)

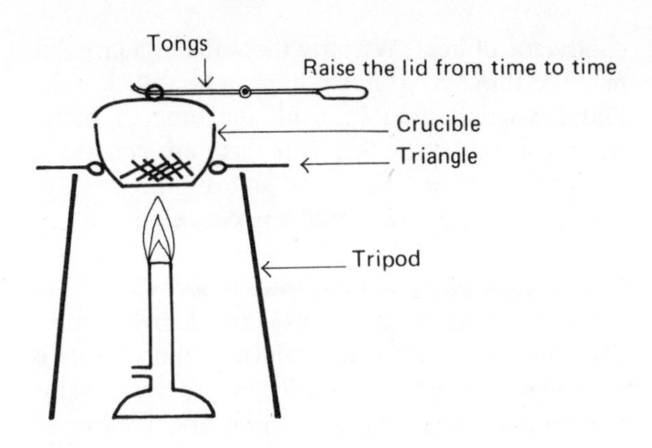

Fig. 16.8. Magnesium – mass and burning.

From time to time lift the lid to see what is happening inside – but do be careful when you do this. After all you do not want chemicals to escape. When you think that all the magnesium has been used up, *let the crucible cool down* **and then weigh it again.**

Mass of crucible, lid, magnesium	=	**g**
Mass afterwards	=	**g**

Crucibles are made of porcelain and this is a poor

Fig. 16.7. The inside of a haystack may get hot because the grass undergoes chemical changes as it dries out. The chemical energy changes into heat energy. Why is a stack of hay a good insulator? How would you try to avoid such a disaster if you were a farmer?

conductor of heat. Why are the walls of a crucible made so thin? What might happen to a thick wall? Did the magnesium burn all the time or better when you raised the lid? Was there an increase in mass? Could the change be due to the crucible in any way, and how would you check on this?

Does phosphorus gain mass when it burns?

For this experiment we use an element called phosphorus – what is an element? Phosphorus is a strange element which catches fire very easily. It is so dangerous that we store it under water all the time. Notice how your teacher handles it with tweezers or tongs – it is most unwise to touch it. See what happens when a piece of it is held out of the water for a few minutes.

Take a piece of phosphorus the size of a pea, blot it dry with a filter paper and put it at the bottom of a test tube. Poke glass wool into the tube using a glass rod until the tube is half full. Leave some space over the phosphorus and pack the wool very loosely (see Fig. 16.9). Now weigh the tube on a top pan, protecting the pan with a cork mat. Write down the mass. Check the mass every minute.

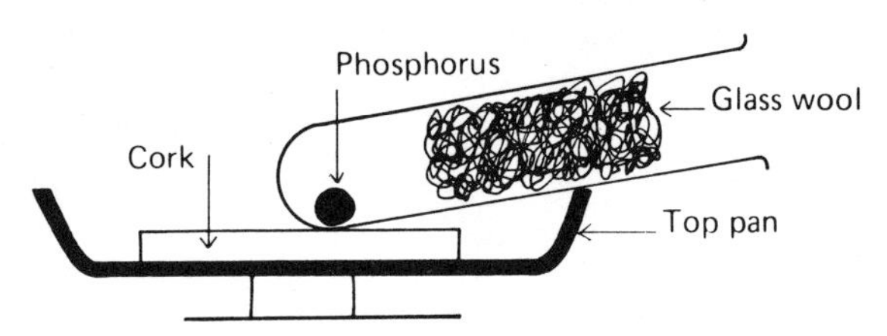

Fig. 16.9. Phosphorus – mass and burning.

Did you see the phosphorus smoking? Did it gain mass?

Both phosphorus and magnesium gain mass when they burn. Here is a puzzle: how can a thing put on mass when you add nothing to it? But if you think hard you will see that something was touching the phosphorus and the magnesium all the time. It was invisible but it does have mass, and a long time ago you did an experiment to show this. What is this something?

Let us make a theory about burning. 'When anything burns it combines with the air to make new chemicals.' Do you think that this explains all the facts you know? Does it explain why things gain mass? Why does a fire burn brighter when you poke it? Sometimes when you stand too close to a fire your clothes catch alight. Did you know that the best thing to do if this happens it to wrap a blanket or a rug round you? How does our theory explain this?

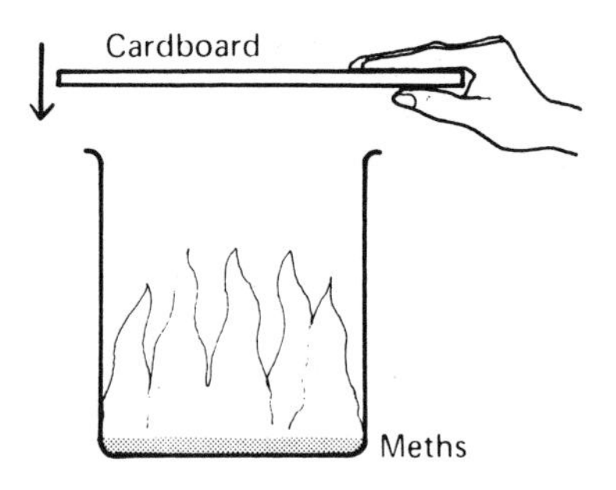

Fig. 16.10. Putting out fires.

Just cover the bottom of a beaker with meths and then put the bottle of meths back on the shelf. Light the meths in the beaker. Blowing will not put the fire out – it may make it worse. But put a piece of stiff cardboard over the mouth of the beaker and leave it there for a few seconds. What happens? Put a small candle in a jar and light it with a taper. Now put the lid on the jar. What happens?

Does our theory explain these things? But like all good theories, ours leads us to ask some more questions. What happens if there is insufficient air? If burning is a chemical reaction what new things are made?

Burning a candle without much air

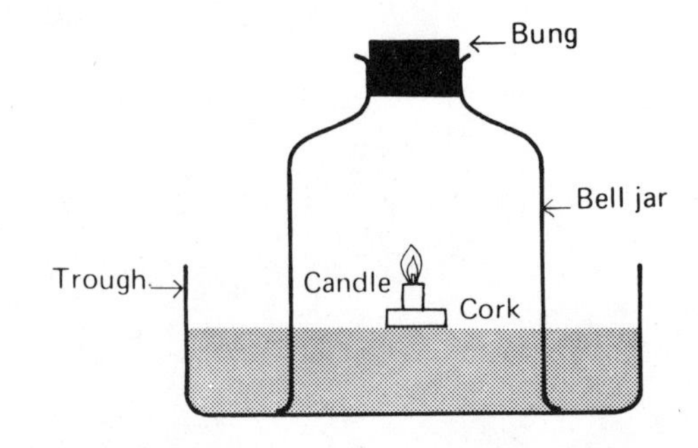

Fig. 16.11. Burning a candle without much air.

The apparatus is shown in Fig. 16.11. A candle is stuck on a cork so that it can float on water. It is then lit, and a big jar (called a bell jar) with its rubber bung taken out is lowered over the candle and the bung is put back.

Now that you have seen what happens you will want to try it again, just to make sure that it happens

every time. Did the candle go out? Was all the air used up? Do you think that air is a mixture? How much of the air was used up – was it a half, a quarter? Judge it roughly.

Burning a piece of phosphorus without much air
You will have noted that a candle uses about $\frac{1}{6}$ or $\frac{1}{7}$ of the air when it burns. Does burning phosphorus do the same?

Use the same apparatus as last time but instead of a cork, float the phosphorus in an evaporating dish. Put the jar over the dish with the bung out. Now heat a long piece of wire till it is red hot at one end and then poke it through the mouth of the jar so that it touches the phosphorus (Fig. 16.12). As soon as the phosphorus catches fire take the wire out and replace the bung.

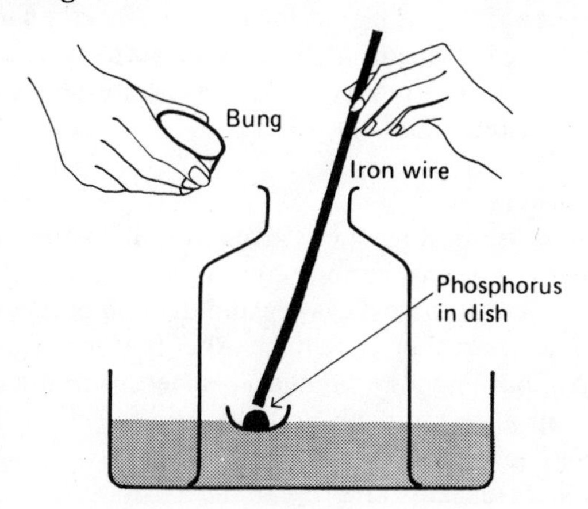

Fig. 16.12. Burning phosphorus without much air.

Watch what happens until the jar is quite cool again. Now answer the following questions.

How much of the air was used up by the phosphorus? Why was it important to let the jar cool down before judging how much air was used? (remember what happens to air and other things when they are heated up). Do you think that both the candle and the phosphorus used up roughly the same fraction of air?

Experiments like the ones we have just done tell us that the air has two different gases in it. One of them is used up when things burn, and the other one seems to have no part in burning. When a chemical takes no part in a reaction we say that it is *inert*. In burning, part of the air is inert, and only a small fraction is used up. I wonder if you have a theory about what this small part is? What gas makes things burn well? (*Clue:* a wooden splint.)

We must now do a test to see if our guess is the right one.

If we burn phosphorus in gas which has been made by mixing air with oxygen, will the phosphorus use up a bigger fraction?

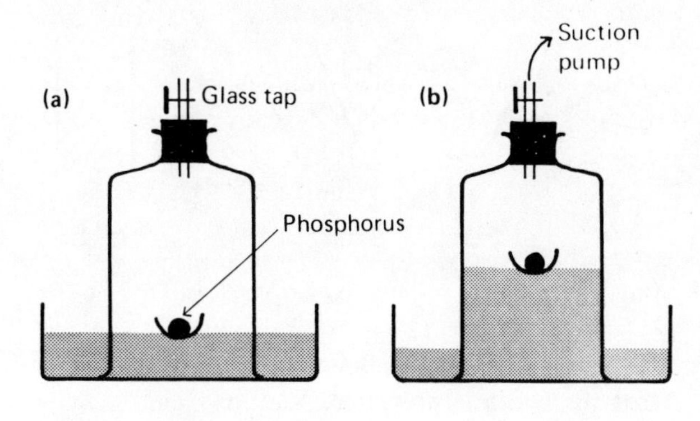

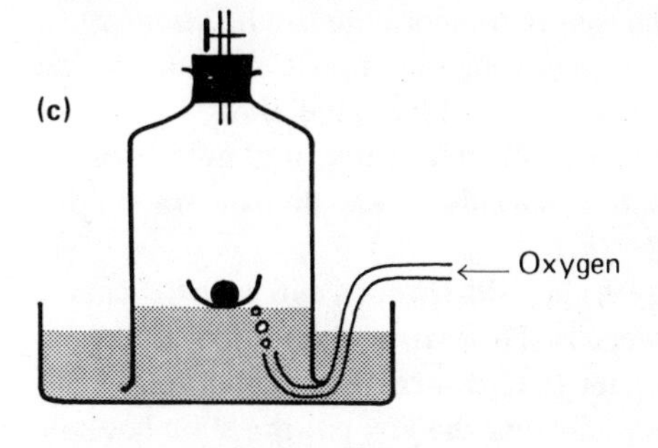

Fig. 16.13. Is the active part oxygen gas?

Figure 16.13 shows how you can try this out. The apparatus is similar to that we used in earlier experiments, but now there is a glass tap through the rubber bung. First make the apparatus as it is shown in (*a*). Next attach a suction pump to the tap and draw out air so that the water is like (*b*). Close the tap now. Bubble oxygen into the jar as shown in (*c*) until the outside and inside water levels are the same. Now light the phosphorus as in the previous experiment and see what fraction is used up.

Experiments like this show that the part of the air which takes part in burning is oxygen gas.

The chemicals made when things burn
What is the name of the chemical made by combining iron and chlorine? Iron and sulphur? Magnesium and oxygen? Because burning things are combining with the oxygen of the air, we expect the substances formed to be called *oxides*. Magnesium burns to give magnesium oxide, sodium would burn to give sodium oxide, etc. It is easy to work out what we will get when an element burns,

but it is a lot more difficult when the substance is not an element. For example, what is made when paper burns, or wax, or wood?

What is made when candle wax burns?

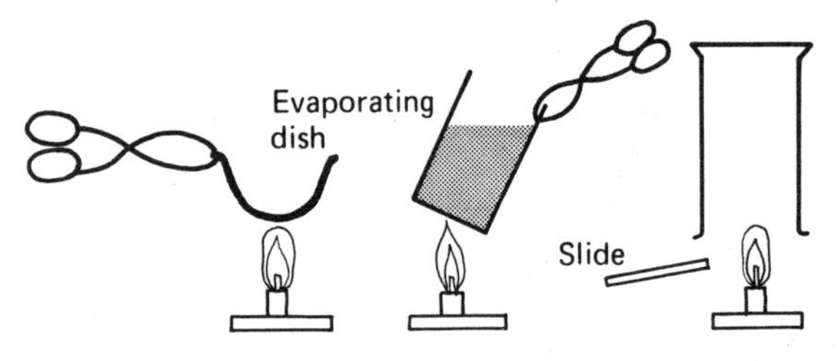

Fig. 16.14.

(1) Stand a lighted candle on an asbestos square so that the bench is protected. You may melt a little wax onto the square to make the candle stand up.
(2) Hold an evaporating dish over the flame so that the tip just touches the white porcelain.
(3) Fill a tin can with cold water, and hold it in the flame for a few seconds. What do you see on the outside of the tin?
(4) Hold a glass jar with its open end over the candle for a few seconds. Now cover the mouth of the jar with a slide and take it away from the flame. Pour 2 cm of limewater into the jar, put the slide back on and shake the jar. What happens? What is limewater a test for?

When a candle burns we get water, carbon dioxide and some soot. The chemist has another name for this last substance, it is called *carbon*. This is an element. Of course the candle may give other things too, we do not know this from our experiments so far.

If you have time do the same experiments on a Bunsen flame. Why not try it with the air hole open and then closed.

You will have noticed that there is condensation on the kitchen walls and windows in cold weather – particularly if you have a gas cooker. You would get this water if the gas was burning but there were no pots and pans on the cooker – why?

A chemical made by burning something else is called a *product of combustion*. Soot, water, and carbon dioxide are the products of combustion of candle wax.

More about molecules

If we know the products of combustion of a fuel we can learn something of what the fuel is made of. For example, what can we say about the molecules in candle wax? Like other molecules they are made of atoms – but what kinds? Carbon is an element and has only one kind of atom in it (what is the symbol for it?). We also know that carbon dioxide has molecules which are made of carbons and oxygens: O—C—O.

Do you think that wax molecules have carbon atoms in them?

Here is a picture of a water molecule:

H—O—H.

What atoms are in it and how many of each are there? Do you think that wax molecules have hydrogen atoms? There is a picture of a wax molecule at the foot of this page.

What a size it is! Count the number of carbon and hydrogen atoms in it. When wax burns there is a reshuffle of atoms to give us the products of combustion. (See Fig. 16.15.)

Summary

Most chemical reactions give out heat energy – we say they are exothermic. Air has at least two gases in it. One of these is oxygen and this combines with things when they burn. The other part is not used up in burning – we say that it is inert. Burning is a chemical reaction which makes new chemicals for us. These products of combustion have oxygen in them (usually) and we call them oxides.

Questions

(1) The product when phosphorus burns in air has molecules:

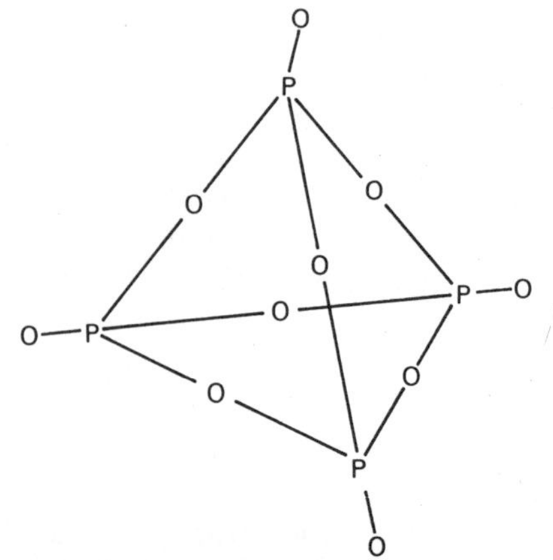

Wax molecule

H H H H H H H H H H H H H H H H H H H
H—C—C—C—C—C—C—C—C—C—C—C—C—C—C—C—C—C—C—C—H
H H H H H H H H H H H H H H H H H H H

where P stands for the phosphorus atom. What is the name of this product? How many oxygen atoms are there in the molecule?

(2) Bismuth, zinc, copper, arsenic, tin, iron, calcium, are all elements. What are the names of the products when each burns in air?

(3) Explain why fanning or blowing a fire makes it burn more brightly.

(4) If a person's clothes catch fire, you must wrap a blanket or a rug round them. Why?

(5) It has been reported that in a forest fire there is always a wind blowing towards the fire. Why is this?

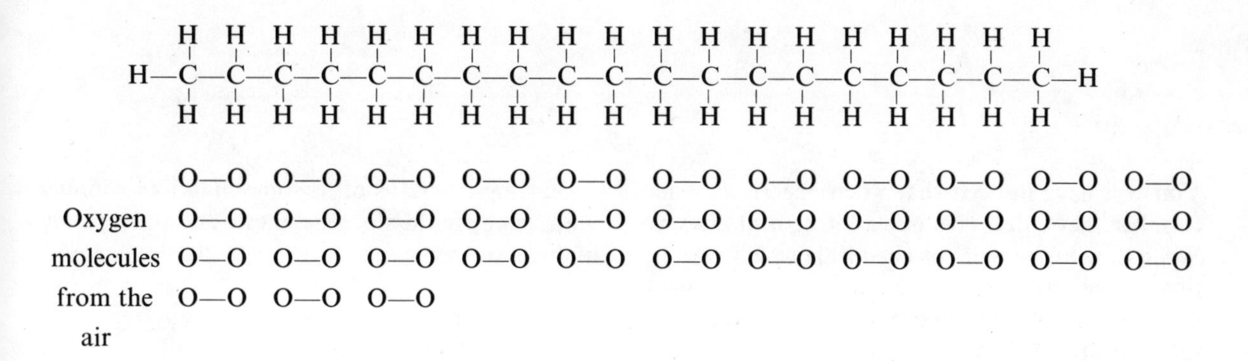

→ Burning

—C—O O—C—O O—C—O O—C—O O—C—O O—C—O O—C—O O—C—O O—C—O
—C—O O—C—O O—C—O O—C—O O—C—O O—C—O O—C—O O—C—O O—C—O
—C—O

H—O—H H—O—H H—O—H H—O—H H—O—H H—O—H H—O—H H—O—H
H—O—H H—O—H H—O—H H—O—H H—O—H H—O—H H—O—H H—O—H
H—O—H H—O—H H—O—H H—O—H

Fig. 16.15. Reshuffling atoms.

Chapter 17 Hurrying Chemical Reactions

You will have noticed that when chemicals react together they often give out a lot of heat energy. Write down four pairs of chemicals which will do this. Some pairs react as soon as you put them together (write down two pairs), but others have to be heated before a reaction will start (write down two pairs). Isn't it strange that we sometimes have to put a little heat in and then we get a lot more out? It is something like firing a gun; you apply a little energy to the trigger and then you get a lot more back! Of course this is a very good thing. If oxygen combined with the coke or coal before we lit them it would be a very dangerous world!

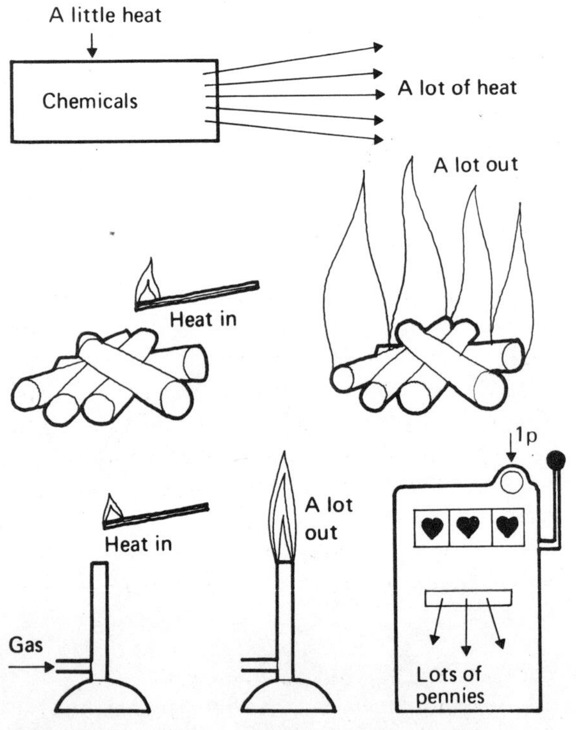

Fig. 17.1. Put in a little and get out a lot.

Starting reactions off is something like winning with 'penny in the slot machines'. You put a penny in and a lot come out.

Can other kinds of energy start reactions?
You may like to see light energy start off a reaction. Hydrogen and chlorine are in the bottle. See what happens when the brilliant light from a flash bulb shines through them. What is the name of the chemical made in this reaction?

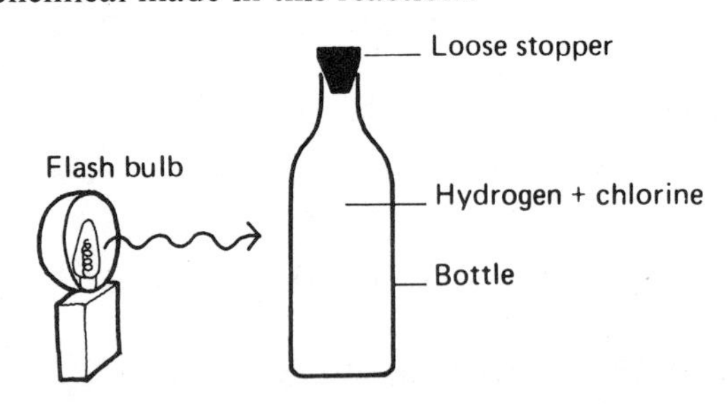

Fig. 17.2. Light starts a reaction.

Will energy make reactions work more quickly?
Some reactions will only work when we give them energy at the start, others seem to go on their own. But what will happen to these others if we put energy into the chemicals by warming them before we start? Do you think they will go faster? Some time ago you discovered that the speed at which sugar will dissolve depends on the temperature of the water. What happens when you raise the temperature? You may now be pretty certain in your own mind that heat energy will speed up a chemical reaction – but we must find out for sure.

You need a 100 cm³ measuring cylinder, some magnesium ribbon, dilute hydrochloric acid, and a thermometer. You also need a watch with a seconds hand.

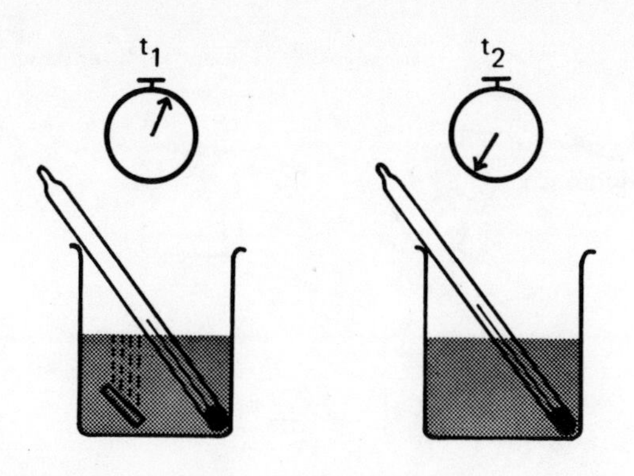

Fig. 17.3. Will energy make reactions go quicker?

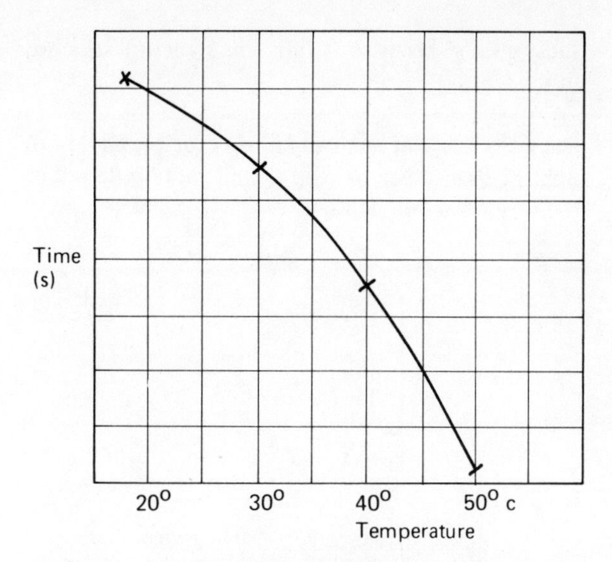

Fig. 17.4. Graph: energy and speed.

Put 100 cm³ of the acid into a beaker. Cut 1 cm of magnesium ribbon, drop it into the acid, and find out how long it takes to dissolve. Stir the acid gently all the time using a glass rod. Read the temperature of the acid at the start and at the end. Now pour the liquid away. Put another 100 cm³ of acid into the beaker, stand it on a tripod and gauze, and warm the acid until it is about 30° C. Take the beaker off the tripod, stand it on the bench, stir the acid and take a careful reading of its temperature. Immediately drop in 1 cm of the ribbon and find out how long it takes to dissolve this time (don't forget to keep stirring it). When it has all dissolved take the temperature again. Do the same thing for 40 and 50° C. A good way to write down your results is:

Time at start t_1 (s)	Time at end t_2 (s)	Temp. start T_1° C	Temp. end T_2° C	Average Temp. $\frac{T_1+T_2}{2}$° C	Time taken (t_2-t_1) s

If you know how to draw graphs it is very interesting to do one putting the temperature along the bottom and the time on the side.

Which acid solution had the most energy and which the least? Which reaction took the shortest time and which the most? Which went fastest?

You may like to try a similar experiment, but this time make acid and marble react together. You will need 4 chips of about the same size (why?). Use 50 cm³ of acid each time, and find the time for a piece to dissolve at 30, 40, 50, and 60° C.

Experiments like this show that reactions go faster when we give them heat energy. Later on (p. 135) we must explain why this happens.

Making reactions faster without using energy

Do you have a bicycle? If so, I hope that you will spend a lot of time looking after it so that it is safe and in good working order. But there are a lot of poor bicycles about and you may have ridden one. Suppose you are riding along at 15 km/hour and you decide to go faster. What must you do? You could put more energy into your pedalling and this would do the trick. But can you think of another way? There is a way of going faster without using more energy – you can oil the works. Do you oil your bicycle every time you use it? The oil seems to last a long time doesn't it; it gets dirty and has to be wiped off, otherwise it would last for a very long time. It is not used up.

We can find chemicals which 'oil' reactions without being used up, and the next experiment shows this.

Starting off methyl alcohol (methanol)

You may not have heard of this chemical, but it is something like meths. In fact we call this common fuel 'meths' because it has some methyl alcohol in it.

Put a few drops of methanol into a crucible and try to light it. Remember to keep the alcohol bottle out of the way when you do this.

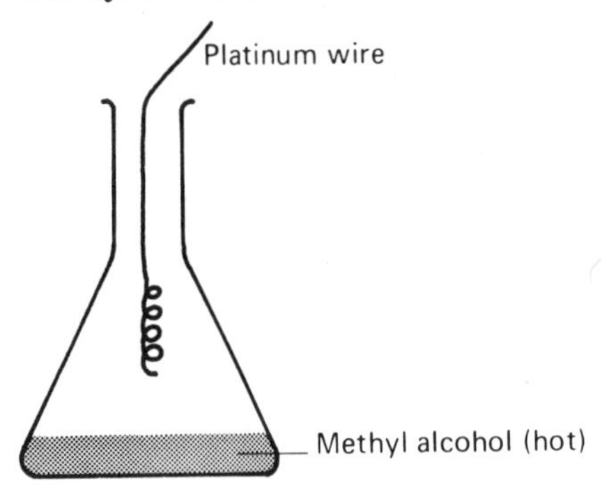

Fig. 17.5. Starting off methanol.

Next, cover the bottom of a conical flask with the alcohol and warm it on a tripod and gauze. When it is almost burning take the Bunsen away. Quickly heat a spiral of platinum wire until it is red hot, and immediately lower it into the conical flask (see Fig. 17.5). Look at the spiral carefully. What do you notice? What is giving the heat energy? Is it coming from the alcohol? Is a chemical reaction taking place? After a minute or so, take the spiral out. Has it been changed?

Decomposing hydrogen peroxide

Pour hydrogen peroxide into a test tube to a depth of 2 cm. Which elements are in this compound? Look at it carefully – is it decomposing to give one of these elements? Stand the tube in a test tube rack and add a spatula of manganese dioxide. What do you see? Test the gas – what is it? Could both chemicals be giving the gas? When no more gas comes off examine the tube. Can you see any new substance? Does the liquid look different? Add some more manganese dioxide. Do you get any more gas? Is the liquid still hydrogen peroxide? Now let us see whether the manganese dioxide has been changed too. If it is just oiling the decomposition of the 'peroxide' it ought to last a long time.

One-quarter fill a test tube with hydrogen peroxide and add a spatula of manganese dioxide to it. Get ready to filter the contents of the tube, and when fizzing has stopped, pour the contents of the tube into the filter paper, collecting the filtrate in a beaker. Pour more hydrogen peroxide into the filter paper. Does it give oxygen now? When this has drained, pour in more hydrogen peroxide. Do this many times.

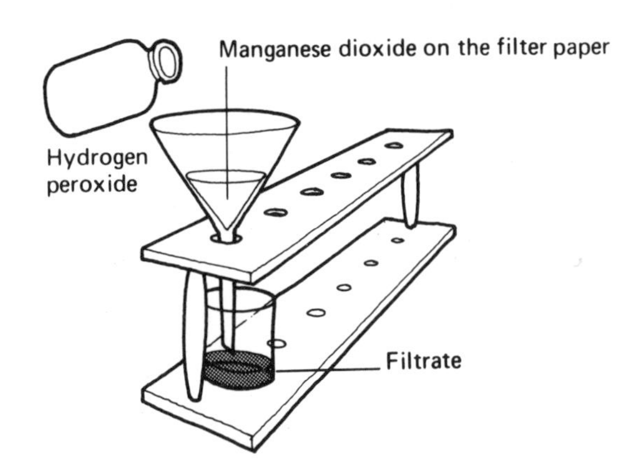

Fig. 17.6. Decomposing hydrogen peroxide.

Do you think that the manganese dioxide is being used up? See if other black powders will do this as well as the manganese dioxide.

Manganese dioxide causes hydrogen peroxide to give off oxygen, but it is not itself used up. A chemical which speeds up a reaction and is not used up is called a *catalyst*. When a reaction has been speeded by a catalyst we say that it has been *catalysed*.

Catalysts are very important in laboratory work and they are equally important in industry. You

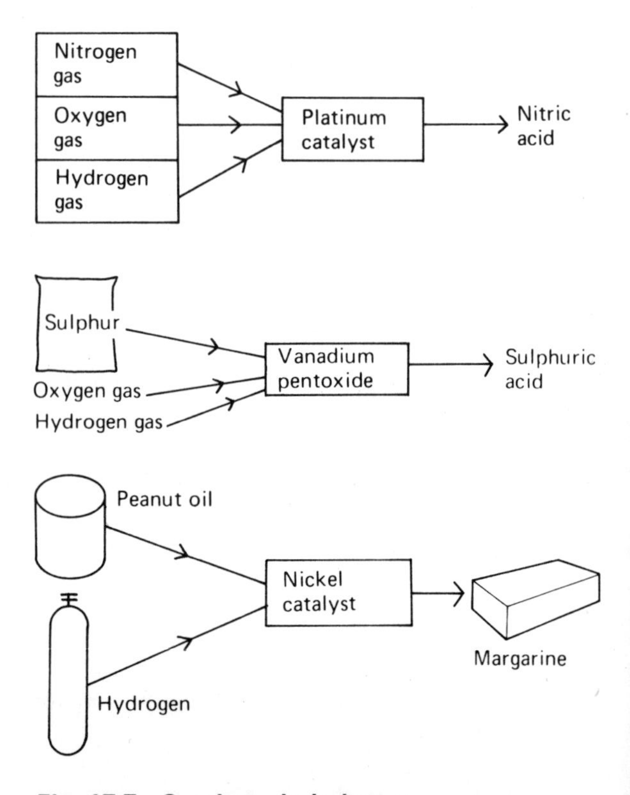

Fig. 17.7. Catalysts in industry.

have used nitric acid, and this important chemical is made by combining its elements (nitrogen, hydrogen and oxygen) using a platinum catalyst. In the same way, sulphuric acid is made from sulphur, hydrogen and oxygen using a catalyst – this time vanadium pentoxide. Margarine is made from edible oils and hydrogen, and these will combine when a nickel catalyst is there. The gas you use in the laboratory could well have been made by reactions which are speeded using catalysts. Oil and steam are mixed together, and the hot mixture is passed over metal catalysts. Plastic things like polythene, synthetic rubber and fibre glass are all made using catalysts. No wonder that people spend a lot of money trying to find new and better ones.

Not all catalysts are as expensive as platinum, but they all cost a lot, and a lot of trouble is taken to keep them in working order. If they get dirty they stop working properly, and we say that they are *poisoned*.

More about molecules

Why is it that energy starts reactions off and makes them go faster? Have you ever played musical games where you dance about the room to music, going faster as the music goes faster, and slower as the music slows down? When the music is fast the game becomes rough because of all the hard bumping with each other. When we heat things up what happens to their molecules? Will they bump into each other harder and more often?

When coal gas and air are mixed up, molecules of gas and air are moving about and hitting each other. But they just bounce off one another and so no new molecules are made. When the gases are heated – a match is put near – the molecules move very quickly. They bang into each other with such force that they come apart, and new molecules (products of combustion) are formed. A chemical reaction has taken place.

Hydrogen molecule
Cold
slow
Oxygen molecule
hit
No change
Hot
fast
hit
H—O—H
Water molecule
New things made

Fig. 17.8. Hydrogen and oxygen combine.

How catalysts work

If you want to stick two balls together you have trouble because they tend to roll apart before the glue is dry. If you want to make a model of an oxygen molecule, it will be worth your while to make two hollows in a piece of wood or plasticine so that the balls stay in place until the glue is set.

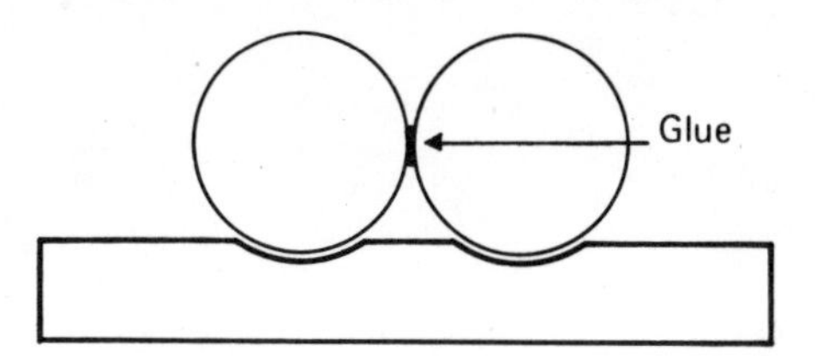

Fig. 17.9. How catalysts work.

When the molecules hit each other they can only make new molecules if they stay together for long enough, and this is where the catalyst comes in.

The surface of the catalyst has got humps and hollows just the right size to fit molecules. The molecules to be reacted settle into the hollows and they stay there long enough for new links to be made. When this has happened the new molecule leaves the surface. What do you think happens when a catalyst is poisoned by dirt? Could it do its work if the hollows were full of dirt?

Summary

Some chemical changes happen without heating, but others need to be heated. All reactions go faster when the temperature is raised, and this is because the molecules move faster and strike each other harder. Sometimes we can make a reaction go faster if we add a catalyst. This speeds up the change but is not changed itself.

Something to do at home

You may have heard that sugar gives you energy. Well you certainly have energy because you can change things. When you get into bed you warm it up, and you change its shape energy. Your energy comes from the chemical reaction of sugar and oxygen inside your body. It is true to say that the sugar is slowly burning away, and when it is used up you feel hungry and tired and need to eat some more.

But now try to burn a lump of sugar in the air.

Put a lump on a saucer and try to light it with a match. Now rub the corner of the sugar in some cigarette ash and try to light it again – putting the match near the ash. Does it light now? Would you say that the ash is a catalyst? How do you know that the ash will not burn? Inside our bodies sugar and air react even though they are quite cool. Special catalysts make this happen.

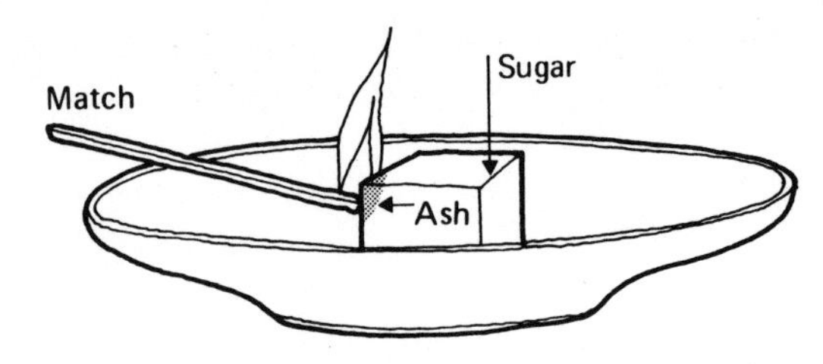

Fig. 17.10. Burning sugar.

Chapter 18 **Flames**

Why do some flames give out light?
What is a luminous flame? How do you make a Bunsen give this kind of flame? When a Bunsen is like this it looks something like a candle, doesn't it? Let us see if both these flames give out light for the same reason. What happens when we hold a clean evaporating dish in these flames?

Make your Bunsen give a small non-luminous flame and hold a clean dish just above it. Is this a sooty flame?

We can now make a theory about why some flames give out light; it is because they have soot particles in them. On the other hand, clean flames have no soot in them and so they are not luminous. Let us test this idea.

Make your Bunsen give a large roaring flame. Sprinkle some soot onto a piece of paper and hold the paper near the flame. Gently blow the soot into the flame (Fig. 18.1). What do you see?

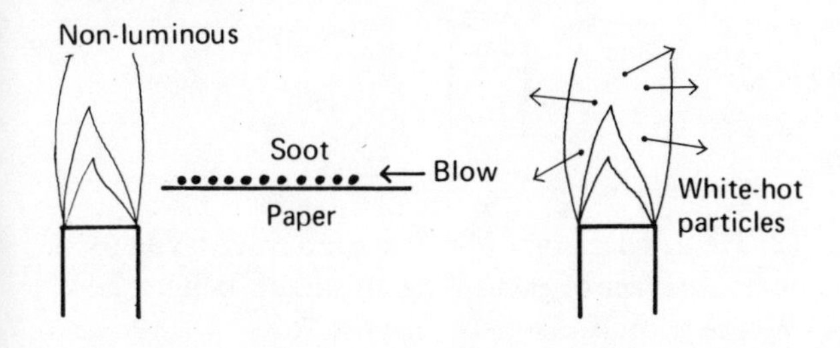

Fig. 18.1. What makes a flame luminous?

Soot is an element – what is its other name? Inside the flame the heat of the reaction makes the soot particles very hot so that they shine white hot. In a clean flame there are no particles to glow.

What is happening in the dark part of a flame?
If you look at a flame you will see that some parts are quite dark. This is very easy to see if you use a Bunsen with the air hole open.

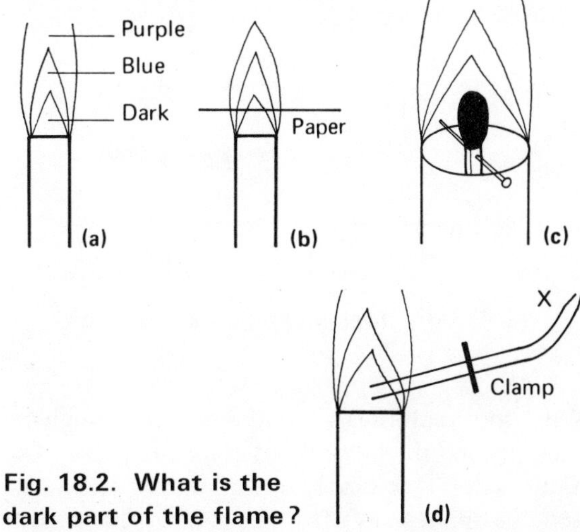

Fig. 18.2. What is the dark part of the flame?

Light your Bunsen and look at the region just above the barrel (Fig. 18.2 (*a*)). Could it be that the burning has not started yet? What would you expect to find if this is true? Let us do some experiments to check up on this. Move a piece of filter paper into the flame just above the barrel. Hold it there for a fraction of a second and then move it away so that it does not catch fire. What is the scorch mark like (Fig. 18.2 (*b*))?

Stick a pin through a match just below the phosphorus end. Lower the match into the barrel of an unlit Bunsen so that the pin rests on the barrel and the match end is held just above it. Open the air hole, turn the gas full on and light the Bunsen (Fig. 18.2 (*c*)).

Clamp the special tube shown in Fig. 18.2 (*d*) into position, light the Bunsen, and then try to light the tube at X.

These experiments show that the dark part of a flame is made of unburnt gas, and that it is fairly cold. Now you can avoid the common mistake of putting the thing you are heating as near to the barrel as possible. Remember that it is not even hot enough to light a match there. Where is the hottest point?

Why must a candle have a wick?

Have you ever tried to light a piece of candle wax without a wick? Try it and see what happens – but remember to clean up afterwards!

Light a candle and then blow it out. Look at the wick as soon as you have done this. What do you see happening? What happens when you heat solid wax? What happens if you heat the liquid wax? The smoke you saw was wax gas.

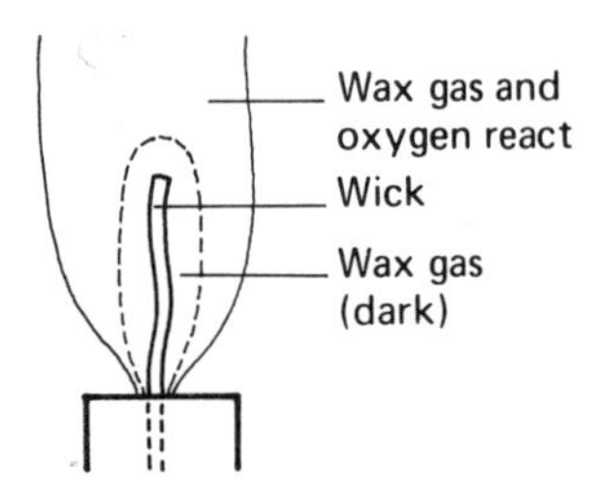

Fig. 18.3. Why does a candle need a wick?

When the candle burns, liquid wax rises through the wick just as ink will soak into blotting paper. The liquid wax is then heated up and turned into a gas, and this diffuses away from the wick. What experiments did you do to show that gases diffuse? Whilst the wax gas diffuses out from the wick, air mixes with it and the gas then burns (what are the products of combustion?). When you try to light wax without a wick, you melt the wax but it runs away before you can make a gas from it.

Summary

Flames are spaces where gases are chemically combining. Even when wood or coal is burning, it is a gas from the fuel which is combining with the oxygen of the air. The dark part of a flame is the fuel gas before it has caught fire, and this is the coolest part of the flame. The luminous part is due to tiny soot particles which are glowing white hot.

Fig. 18.4. This is the flame of an oxy-acetylene welder. The oxygen and acetylene come out of the nozzle and burn to give a very hot flame (3000° C). It can melt iron and most other metals. Go to your local garage and find out whether they can do this kind of welding.

Something to do at home

Sketch the flame of a match, a taper, a spirit lamp or a cigarette lighter. Label the parts that you think are:

unburnt gas, very sooty, hottest.

Questions

(1) Gas used to be used for lighting purposes, and this made ceilings get very dirty. Explain why.

Fig. 18.5.

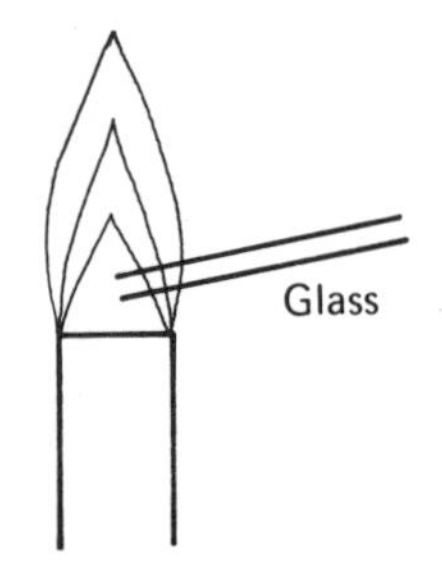

(2) Here is a picture of a Bunsen burner being used to heat a piece of glass. What mistake is being made? Where should the piece of glass be?

Chapter 19 Electrical Energy

In learning about energy, we discovered that there are many different kinds of energy, and that one kind can change into another. One of our experiments was to use the place energy in a weight to turn a dynamo and to use the dynamo to light a lamp. Write down the energy diagram for this experiment (see p. 98).

If you live in Britain you will all have seen an electric fire and electric light. If you live in a hot country you may not have seen an electric fire, but you will probably have seen an electric torch and perhaps an electric fan. Some people have electric cookers or washing machines. Electric lamps and fires turn electrical energy into heat and light; fans and washing machines change it into motion energy. If you have used any of these you will know that you have to plug in to the mains if you want them to work. You may have used an electric torch, and if so you will know that the bulb must be properly screwed in and a battery or cell supplied before you can hope to get light from it. When you switch on, the battery's chemical energy changes into electrical energy, and this turns to heat and light energy in the lamp.

We are now going to find out more about this electrical energy and what happens when we switch on. We shall use small bulbs and batteries for these experiments because the mains supply a great deal of energy – enough to hurt you or even kill you if you were careless. Never use the mains supply for any electrical experiments at home.

Experiment: *How can we light the bulb?*

Look closely at the little bulb you have been given. You will see that inside the glass there is a very fine wire that gets very hot when the bulb lights up. This is called the *filament*. At the bottom of the glass bulb there is a metal screw and below that a soft metal stud. Take the bulb and hold it so that its stud touches the battery stud as in Fig. 19.1 (*a*). Does the bulb light up? Try holding it as in Fig. 19.1 (*b*). Does the bulb light this time?

Fig. 19.1.

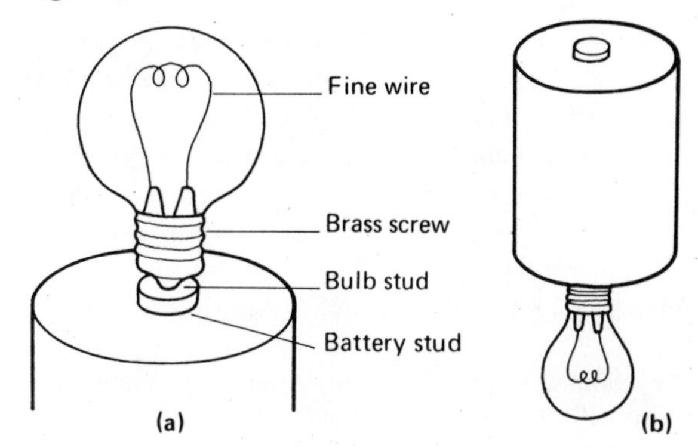

Hold the bulb once again as in Fig. 19.1 (*a*). Your partner has one length of copper wire. See if, between you, you can light the bulb using only the piece of wire and the battery. Now hold the bulb as in Fig. 19.1 (*b*) and again use the wire to light the lamp. The battery stud and the bottom of its case are the two ends of the battery – scientists call them the *poles* of the battery. (I expect you know that the Earth's north and south poles are at opposite ends of it.)

Try the experiment again using other things besides copper wire, for instance (*a*) a piece of string, (*b*) a bent piece of tin or aluminium or brass, (*c*) a piece of cardboard, (*d*) a piece of wool, (*e*) a bent twig, (*f*) a piece of raffia. You may be able to think of other things to try.

We call the substances that allow the bulb to light *conductors* and the others *insulators*. Write down your conclusion by copying into your book the next sentence and filling in the gaps:

To make the bulb light we have to connect the stud of the bulb to one .... of the battery and then join the bulb screw to the other .... of the battery using a .... (usually a piece of ....).

Fig. 19.2.

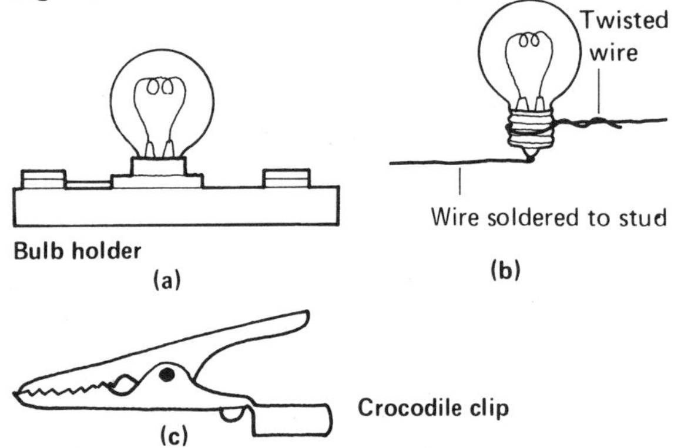

Look carefully at the bulb and holder given to you before you screw the bulb in. How does it hold the bulb? What touches the bulb stud when the bulb is in place? On the holder there are two screws to which wires can be fixed. Are they connected by pieces of metal to the bulb screw and bulb stud? Why? What will you expect to happen if you screw the bulb in position and then connect the two screws to the two poles of your battery? Try it. If your bulb is mounted as in Fig. 19.2 (*b*) you can see at once how to light it.

Experiment: *More experiments with batteries and bulbs*

You will require some croc leads. These are pieces of wire with crocodile clips (see Fig. 19.2 (*c*)) fixed to their ends. Why are they called crocodile clips? Connect the bulb (in its holder if you have one) to the cell using the croc leads and notice how it lights up.

Fig. 19.3.

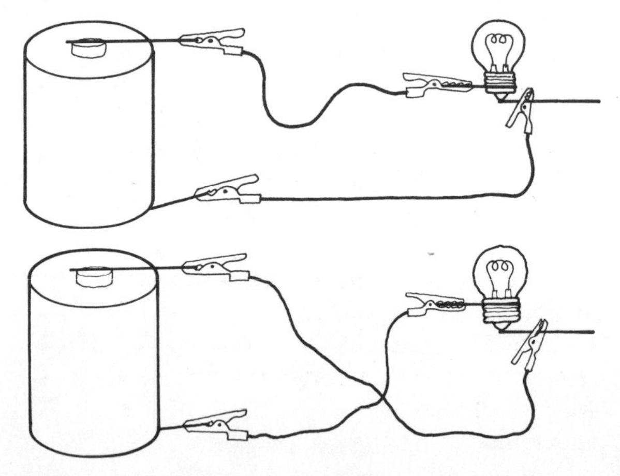

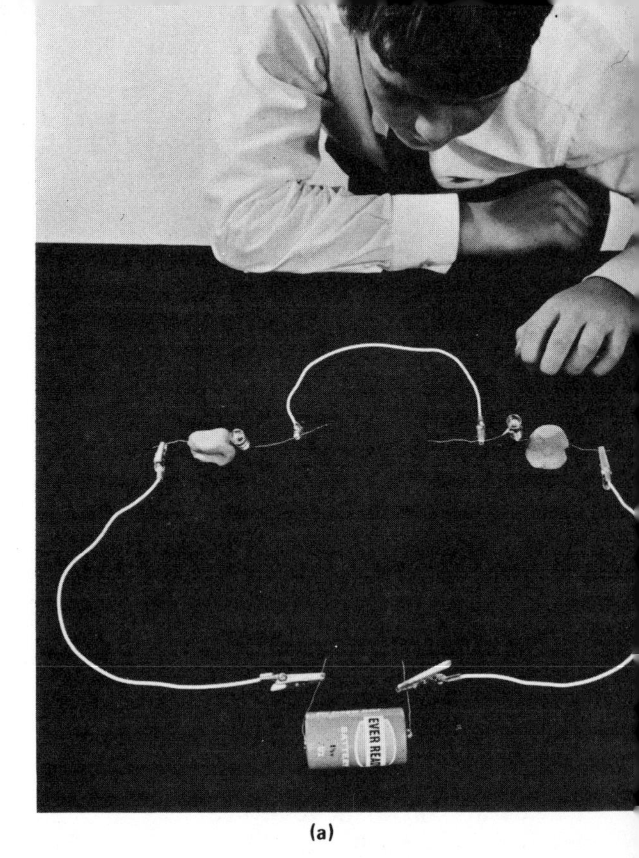

(a)

Fig. 19.4. Lighting two bulbs from one cell. Will the bulbs light up as brightly as they should? in (*a*)? in (*b*)? In which picture are they arranged in parallel?

(b)

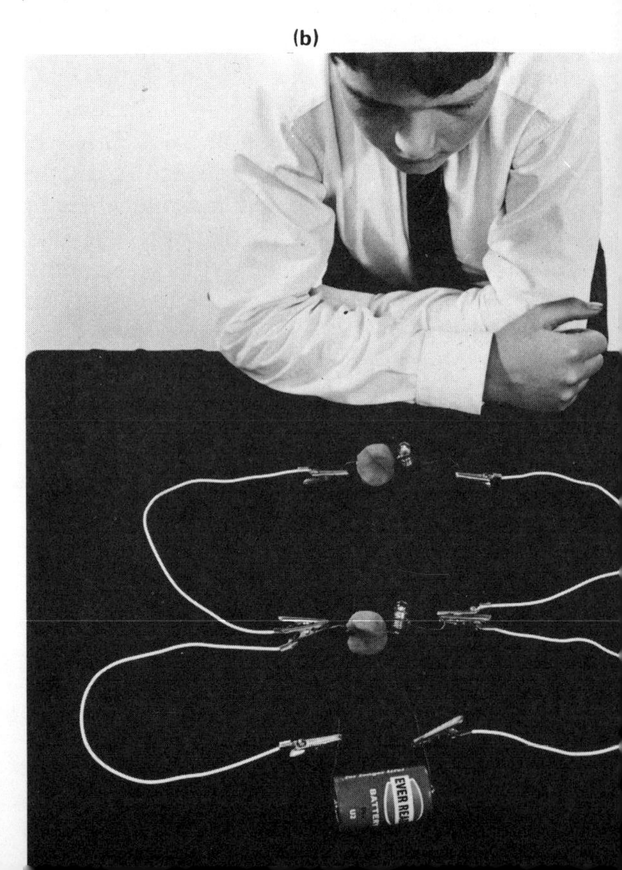

Does it matter which end of the holder you connect to the stud and which to the bottom of the cell? (Fig. 19.3).

Can a cell light two bulbs?
Take two bulbs and connect them to the same cell. There are two ways of doing this – think out how the bulbs can be arranged and then try the arrangements. You will need extra croc leads – ask for them. Look carefully at the bulbs and see how brightly they shine each time. Do they shine as brightly as the single bulb? Draw in your book sketches like those in Fig. 19.3 to show how you lit two bulbs from the same cell.

When you have drawn these sketches you may have decided that it takes a long time to draw sketches of electrical apparatus. Electricians think this too, and so that they can work more quickly they use diagrams. You remember that a long while ago we showed you the difference between a sketch or drawing and a diagram (p. 5). A diagram usually looks something like what it stands for, and electricians sometimes use diagrams of this kind, but more often they use *symbols*. Symbols don't usually look like the things they stand for. You have met some before. You know that the symbol K stands for an atom of potassium, but we don't think that an atom *looks* like a letter! Electricians use symbols to stand for pieces of apparatus and they use ⊣⊢ or ⊣|⊢ to stand for a cell, although it doesn't look like one. You can imagine that the long stroke stands for the battery stud and the short one for the bottom of the case. In electrical diagrams connecting wires are shown just as lines, and we shall use the sign B to stand for a lamp bulb. So the electrical diagram for a battery connected to a bulb will be that shown in Fig. 19.5.

Fig. 19.5.

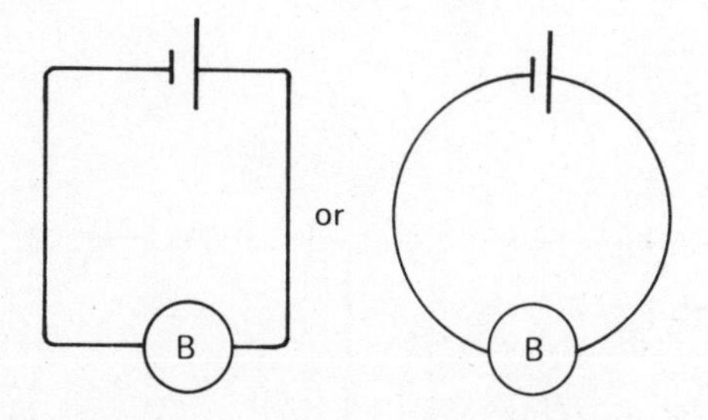

Does it matter if the wires are straight or curved? Which way looks neater in your book?

We call a diagram like this a *circuit diagram*.

Draw circuit diagrams to show how you connected the one battery to the two lamps.

When talking about flash lamps some people use the word cell and some the word battery. You can use either, and we shall sometimes use one and sometimes the other. There is a slight difference in meaning but you need not bother about it now.

Experiment: *What happens if we try to use two cells to light one bulb?*

Before we can do this we must decide how to join the two cells together. Two possible ways are shown in Fig. 19.6.

Fig. 19.6.

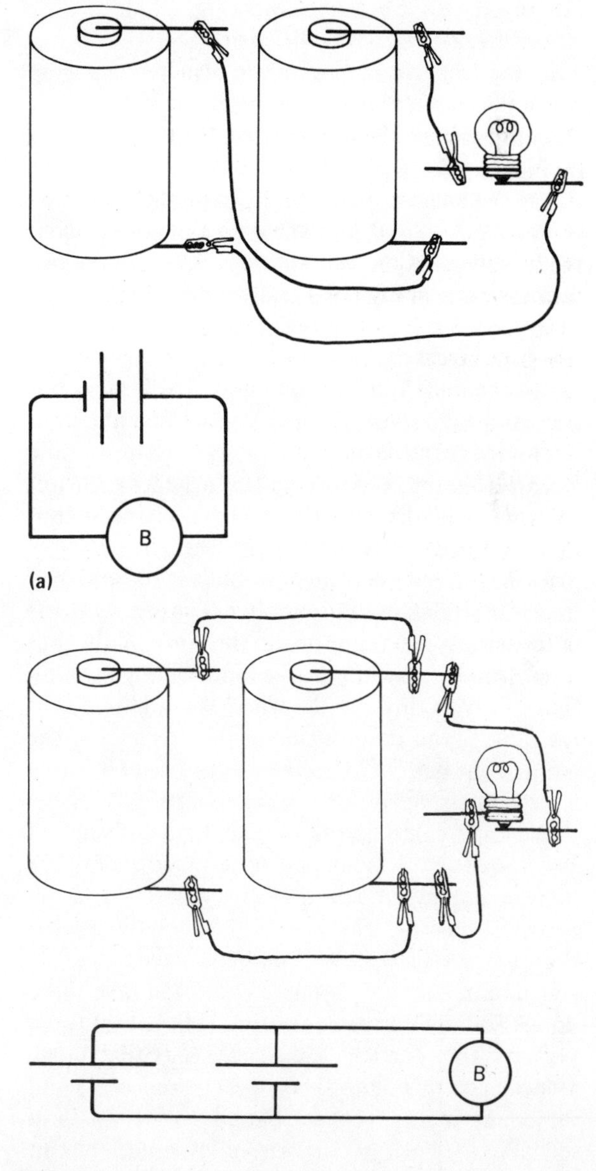

Try them. You may be able to think of other ways of joining the two cells to the bulb. If so, try them, and draw circuit diagrams.

The bulb may light up normally (as it did when connected as in Fig. 19.3) or more brightly than this, or less brightly, or not at all. Under your diagrams, and also under those you drew to show one battery connected to two bulbs, write Normal, Brighter, Dull or None, to show how much light the bulb gave in each experiment.

Experiments

(1) Arrange your bulb and batteries as in Fig. 19.7 and notice whether the *order* in which you put bulbs and batteries matters.

(2) Now try these arrangements (Fig. 19.8).

Are the last two really different?

In the last five arrangements also try changing round the connections to one battery as in Fig. 19.3. Does this change the way the bulb lights up?

(3) Now try these (Fig. 19.9).

Are the bulbs in Fig. 19.8 lit as brightly as those in Fig. 19.7 (*a*) and (*b*)? What do you notice about the brightness of the bulbs in Fig. 19.9? Are the two arrangements in Fig. 19.9 really different?

Fig. 19.7.

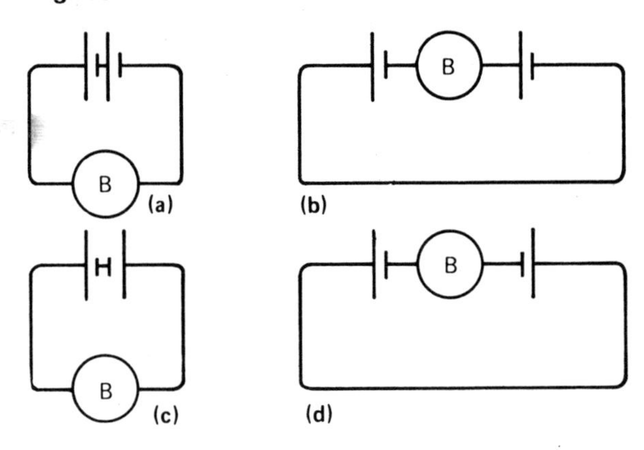

Fig. 19.8.

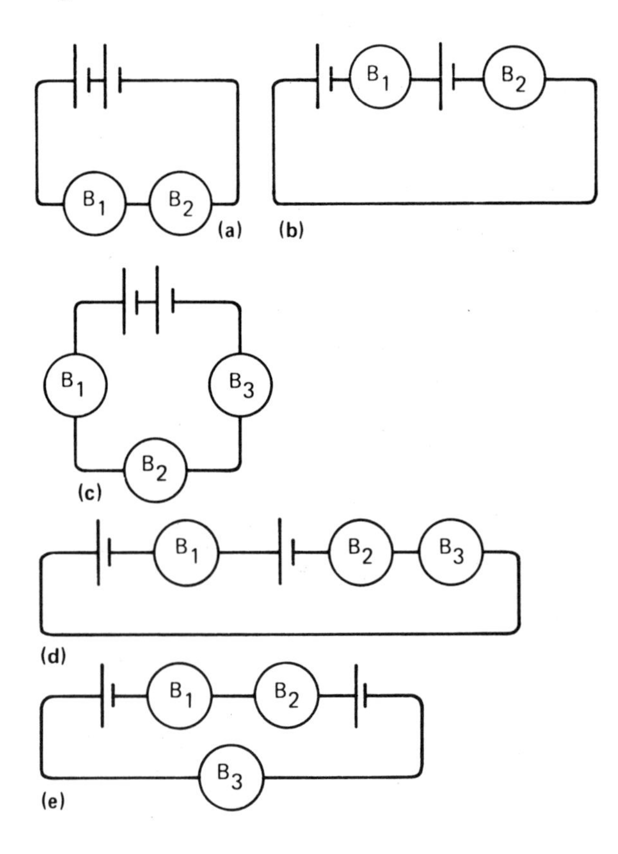

Fig. 19.9.

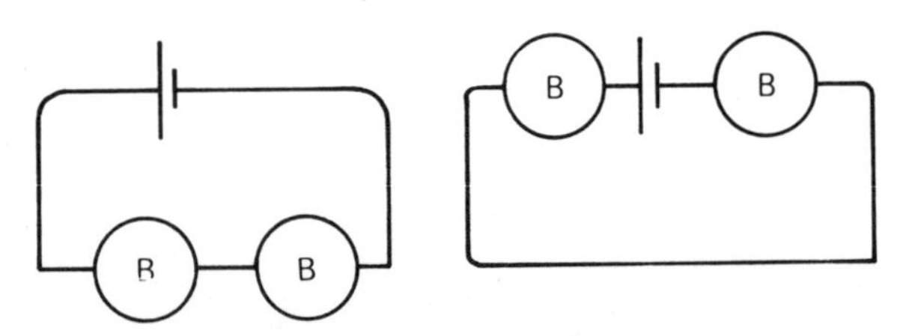

Electric current

When the bulb lights up we know that it gives out heat and light energy. Since it doesn't light until it is joined to the battery the energy must have come from the battery, and we decided earlier that chemical energy in the battery has been changed into light and heat in the bulb. Chemical changes take place in the battery while the bulb is lit, and you may learn more about these later. The energy starts in the battery but it appears in the lamp. How does it get from the battery to the lamp? Do you think that it travels through the wire? What happens to the light if you remove one of the crocs from the lamp or the battery? Can the energy travel through a piece of string? a piece of cardboard?

We have come across various kinds of energy. We know that a stone can have place energy, or motion energy, a spring shape energy and oil chemical energy. So energy is something that things have, just as you perhaps have a pen, a pencil, a handbag or a shirt. You often take these things about with you. Scientists believe that there are very tiny particles called *electrons* that move along the wire when it is properly connected, and they carry energy from the battery to the bulb and so make it light up. The wire allows electrons to flow through it as a pipe allows water to flow through it, and we call this flow of electrons a current of electricity or an electric current.

Fig. 19.10.

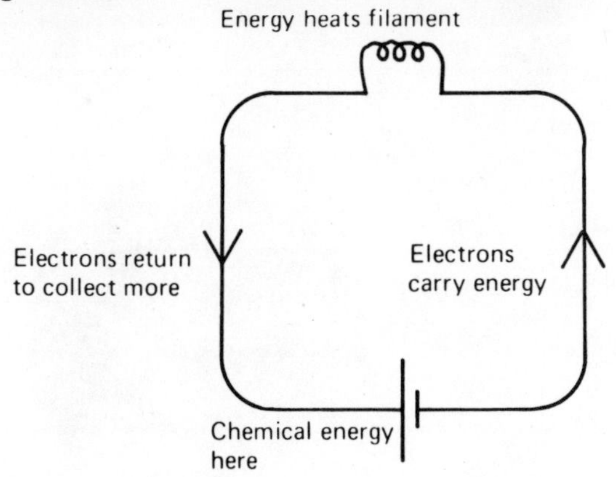

We can't prove to you now that there really are little particles called electrons, and for many years scientists did not know what it was that carried the energy. They thought it was some kind of liquid, and even now electricians sometimes use 'juice' as a sort of slang for current. The simple experiments that we do certainly do not prove that there are electrons moving in the wire. However, when you continue to study electricity and are older you will learn that the results of many experiments make us pretty sure that the energy in a wire *is* carried by electrons.

The biggest difference between water running along a pipe and electrons running along a wire is that if you cut the pipe, water runs out of the cut end but if you cut one of your electric wires or undo a croc clip the electrons *stop* running. This is because water can run easily through air but electrons can't, so cutting an electric wire or disconnecting a croc clip is like turning a water tap *off* – the electrons can't go on moving along the wire. You found that this happened when you disconnected your croc lead – the bulb went out because the electrons could no longer move round and so the bulb got no energy.

What the battery does

You may remember that to get a trolley on the move when it was standing on the table we had to apply a force to it. To get electrons on the move something like a force is needed and the battery or cell provides this. Because it moves electrons we call it an *electromotive force*, and we write it for short as e.m.f. using its initials and small letters.

The electric circuit

Experiment

Use two cells and three bulbs and arrange them like

Fig. 19.11.

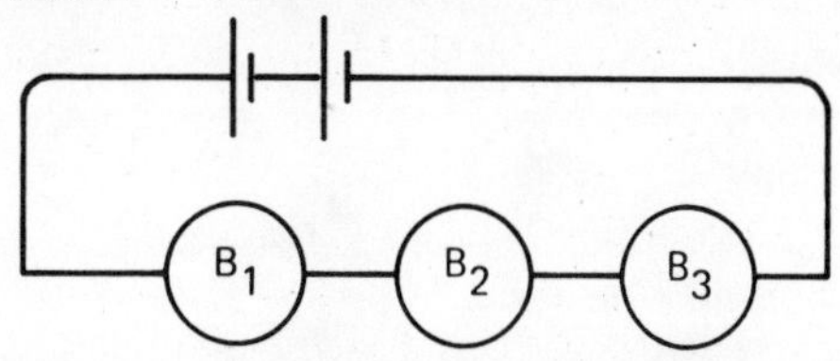

this (Fig. 19.11), using your croc leads as before. Now disconnect the wires, first between the cells and B_1, then between the cells and B_3. What happens to the bulbs? Now try disconnecting between B_1 and B_2, then between B_2 and B_3 and between the two cells. Do the bulbs all do the same thing every time?

This experiment shows that the electrons can't flow unless they can start from one pole of the battery and find a complete path round to the other pole. If we disconnect a wire *anywhere* in their path the flow of electrons stops, the energy no longer travels to the bulbs and they don't light up. The battery is a sort of pump and it provides an e.m.f. that drives electrons round and round. This is why electricians call the path of the electrons a *circuit*. You have seen motor races or athletic races in which competitors go round and round a circuit in the same sort of way.

If you have played with a model railway you can picture electrical circuits like this (Fig. 19.12).

Suppose you have a circle of rails and you fill up the circle completely with trucks coupled together. Now with your hand push the trucks close to you gently in one direction. All the trucks will move round together. If you push a bit harder (or if a friend helps) you'll be able to move them more quickly. We could say you provide a 'locomotive force'. If you pushed in the other direction the trucks would go just as easily the other way. If your friend is helping you, would it matter much whether he sat beside you, or opposite you as in Fig. 19.12, or somewhere in between? What would happen if you pushed the trucks one way but your friend pushed the other way?

You and your friend in the train circuit are like the cells in the electric circuits in Figs. 19.7 and 19.8. If you arrange a train circuit like this you can see that the same length of truck passes over every joint between the rails every minute, and that you can't stop one truck without stopping all of them. In the electric circuits the same number of electrons goes past each point on the circuit and if you stop the electrons passing one place they stop everywhere.

It is not that electrons look like trucks only smaller, but thinking about trucks that you can see

Fig. 19.12. 'Locomotive force'. When the boys push all the trucks move at once. The same length of train passes each rail joint in any given time – in 1 second, for instance.

may help you to understand electrons that you can't see. Electrons are so tiny that even in the finest wire there are millions of electrons abreast, but on your track you can only put one truck. Scientists often find that it helps them to understand what they *don't* know if they think about what they *do* know, but they have to remember that no two things in the world are exactly alike – not even twins – and so they always test their thinking by doing more experiments.

Strip moves down to complete circuit

Springy metal

End view

Wood

Wood

Fig. 19.13. A knife switch. The shaded parts must be made of metal. Why?

Switches

We don't want our flash lamp to give us energy all the time. It would be wasteful to use the energy in bright sunshine, for instance. We can stop the bulb giving light energy, as we have seen, by removing a croc clip. But it would be inconvenient if we had to undo a croc clip every time – and dangerous too if we were working with the electric mains. So we use things called *switches* that are specially made to do the job safely and quickly.

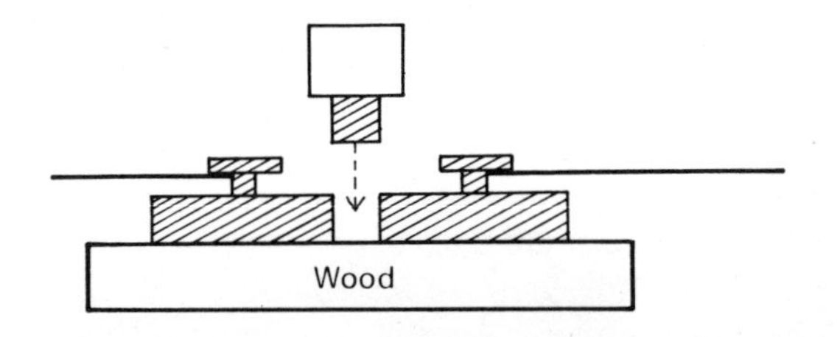

Fig. 19.14. A plug switch. Why is the base made of wood?

Fig. 19.15.

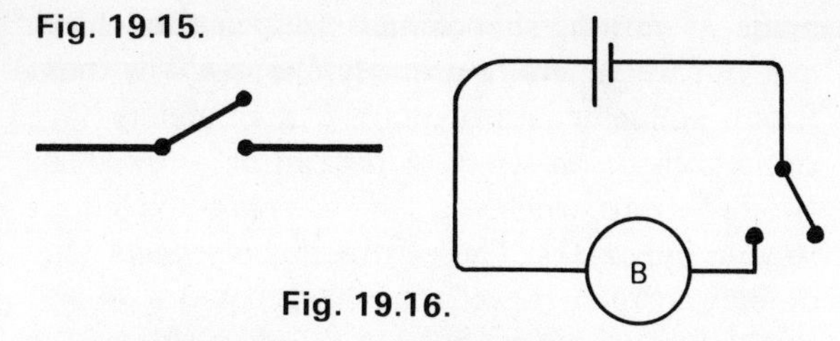

Fig. 19.16.

A symbol that is often used to stand for a switch is this (Fig. 19.15), which represents the switch when it is off, because there is a gap between the two dots. So a bulb in a circuit with a switch would be drawn as in Fig. 19.16.

Question

(1) Draw again the circuit of Fig. 19.8 (*d*). Suppose you wished to put a switch into the circuit. Draw four diagrams showing the switch in four possible places. Which of them do you think is best? Try them all and see.

Which way round?

I expect you have been wondering which way round the current flows and perhaps some of you know – or think you do! When scientists first began to study electric currents they really had no way of knowing which way the electricity flowed, so they just guessed. They called one end of their battery *positive* (+ve) and the other *negative* (−ve) and they said the current flows from +ve to −ve. On our batteries the stud is the positive (long stroke) and the case is negative (short stroke) so the current flows from stud to case. Thus we can now draw our diagram as in Fig. 19.17. We label the +ve and −ve poles and the arrow shows the direction in which they guessed the current flowed.

Fig. 19.17.

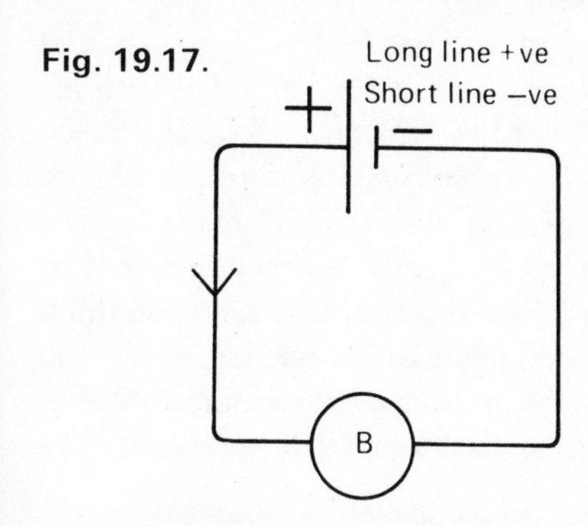

Unfortunately they guessed wrong! Much later scientists discovered that *in a metal wire* electrons flow the opposite way to this. We can't blame the early scientists because electrons hadn't been discovered in their time. By the time they were discovered people had been writing currents this way round for many years and they couldn't be bothered to change their ways. So we still put the arrows showing current flowing from +ve to −ve but we remember that *in a piece of wire* what is really happening is that electrons are going the other way. When currents flow through some other substances (do you remember sending current through some liquids earlier on?) we now know that there are other particles on the move, and some of them move the opposite way to the electrons, so that to draw the arrows pointing the same way as the electrons go would not be as helpful as you might suppose.

Questions

(1) Look back at all the diagrams of Figs. 19.7 and 19.8. Copy them in your book and put in arrows to show which way the current flows.

(2) What happened to the bulbs when you arranged them as in Fig. 19.7 (*c*) and (*d*)? Did they light up? If not, you ought not to put any arrows on that circuit.

(3) Do you think that the brightness of the bulb in any one arrangement depends on whether the wires are straight or curved, or long or short?

(4) Working in a bit of a hurry, you screw a bulb into a torch but when you switch on it doesn't light. What would you do to try to find out why? (Think of as many possible faults as you can.)

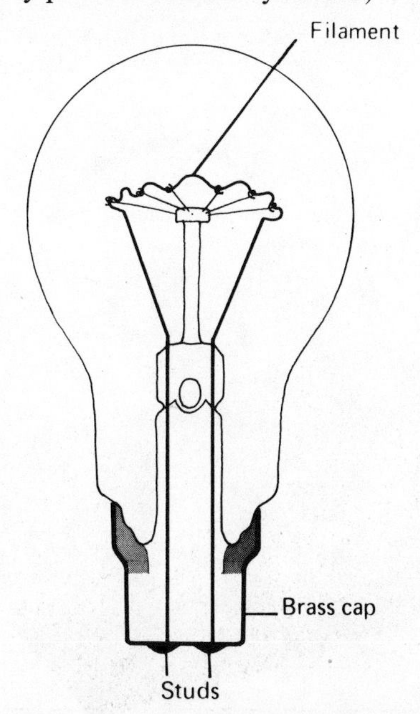

Fig. 19.18. A modern electric light bulb. The filament is made of very fine wire, which is coiled so that it gets very hot.

(5) Find out which scientist invented the electric light bulb.
(6) Who were Ampère, Volta and Faraday? In which countries did they live?

Counting the electrons

So far we have judged whether electrons are going round the circuit quickly or slowly by looking at the bulbs to see whether they light dimly, or normally, or not at all, and so on. This isn't a very convenient way because it isn't easy to judge just how bright a bulb is. So scientists have invented an instrument to tell us how many electrons are going past a place on the wire every second. You need not worry about how the instrument works just at the moment, but it is called an *ammeter*.

Electrons are very tiny, as we said earlier, and so a very big number of them have to travel down the wire if you want to send much energy along it. It would be very inconvenient if you had to write down the actual number because it is so huge, and so scientists measure the number in units called *amperes*. This is why the meter is called an ammeter. The meter reads one ampere when about six million million million (6 000 000 000 000 000 000) electrons go through it every second. Fortunately you don't have to remember this very big number! We can use an ammeter to find out how much current is going through a bulb when it is lit. Ammeters are made in various shapes and sizes, just as clocks are, but we generally use a symbol -Ⓐ- to stand for an ammeter in a diagram. All ammeters have two knobs on to which wires can be screwed, or two holes into which you can push plugs to connect to your apparatus. These are called *terminals*. One of them is brown (or red) and marked with a +, and this you must always join to the +ve side of the battery; the other is blue (or black) and must be joined to the −ve side of the battery.

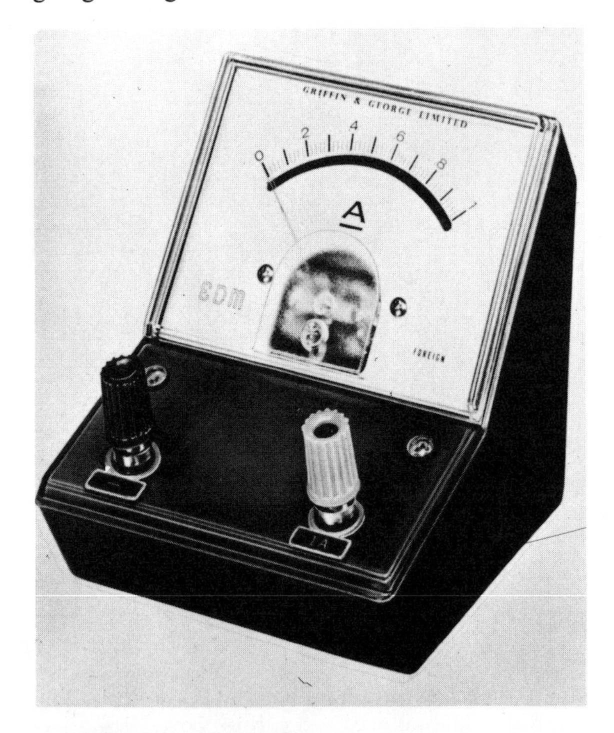

Fig. 19.19. An ammeter.

Experiment: *Using an ammeter*

Arrange a circuit as below, taking great care to join the ammeter the right way round. Record the meter reading. Now put the ammeter between B_1 and B_2 and again record the reading. Try again with the meter between B_1 and the battery, and also with it between the two batteries. Make sure that you screw the wires up tightly each time. Does the meter count the same number of electrons in each position?

Fig. 19.20.

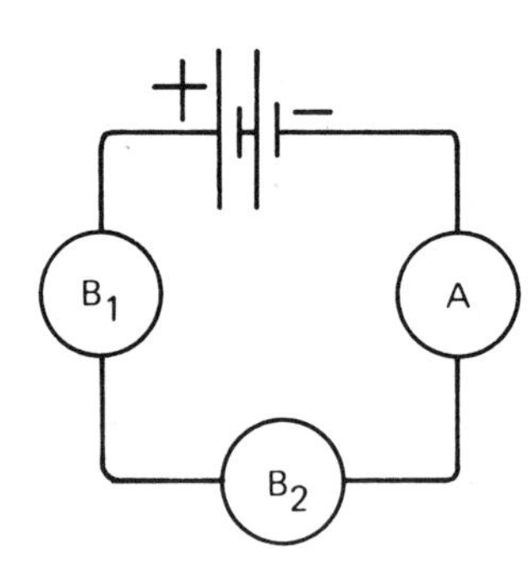

Now remove one bulb – the other bulb will light up extra brightly – and read the current that flows. Make up a circuit so that the current flows from one battery through one bulb and the ammeter and record what current flows.

Make up a circuit with one battery, two bulbs and the ammeter, arranged as in Fig. 19.9, but with the meter as well. Record the current. About how much current does it take to light a bulb normally? What current flows when the bulb is dim? extra bright? Does each bulb need *exactly* the same current to light it normally? Find this out. If not, do you think that you have made mistakes in reading, or do you think that the bulbs are not quite the same? Would you expect bulbs or batteries all to be exactly the same?

When we want to use an ammeter to read the current going through a bulb we must of course make sure that all the electrons going through the bulb also go through the meter to be counted. So we arrange them like this:

Fig. 19.21.

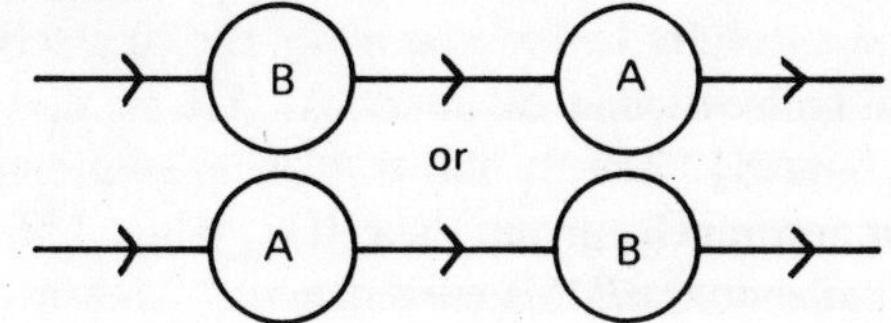

When two pieces of electrical equipment are arranged like this electricians say that they are *in series*. Also electricians refer to 'pieces of electrical equipment' as *components*, because it saves time. Two components are in series if all the electrons passing through either one pass also through the other.

Questions

(1) Look again at Fig. 19.8 (*d*), and answer the following questions:

(*a*) Are all three bulbs in series?

(*b*) Are the two batteries in series?

(*c*) Is one of the batteries in series with B_3?

(2) When two batteries are arranged in series, do they drive more electrons through a bulb than one battery does?

(3) When one battery drives current through two bulbs in series, does it drive more or less current than it can drive through one bulb only?

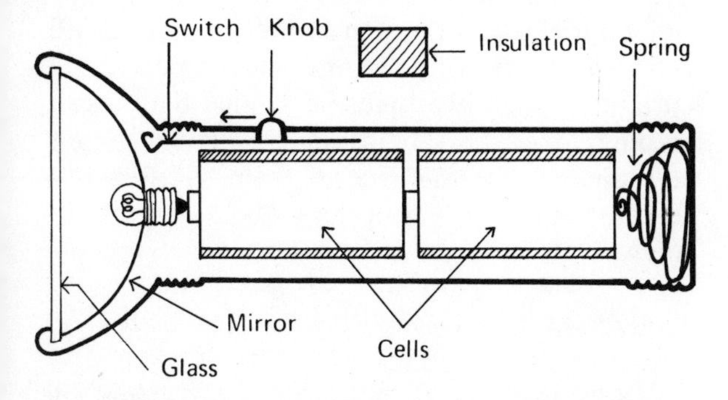

Fig. 19.22. When the knob is pressed in, the circuit is completed and the bulb lights. How does the current travel round the circuit? Are the cells arranged in series or in parallel?

Summary

A battery changes chemical energy into electrical energy. It lights a bulb when energy travels from it to the bulb. In a wire the energy is carried by moving electrons. The battery provides an electromotive force to move them.

The electrons can only flow if there is a completed conducting circuit from one pole of the battery to the other.

The electrons in the wire travel in the opposite direction to the arrow we put on the circuit, which points from positive (+ve) to negative (−ve).

Using two cells in series we get a bigger current through a bulb than we get from one. Using two bulbs in series we get a smaller current from the cell than we do with one bulb.

The current is the same at different places in the circuit and doesn't change if we alter the order of the components.

We can measure the current in amperes by means of an ammeter, which counts how many electrons each second pass a point in the circuit.

More experiments with circuits

Experiment

In the experiment on p. 141 you were asked to find *two* ways of lighting two bulbs from the same battery. One of them is shown in Fig. 19.9. A diagram for the other is shown in Fig. 19.23.

Fig. 19.23.

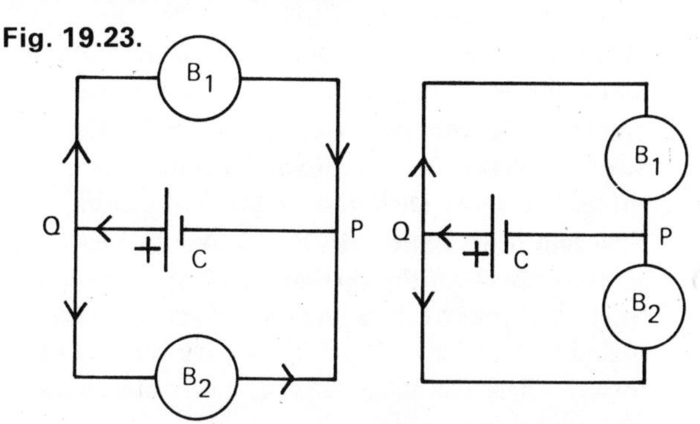

and you may already have discovered this arrangement. If not, wire up this circuit. Do the two diagrams really show different arrangements? Do the bulbs light normally? dimly? extra brightly? Are the bulbs B_1 and B_2 in series?

Although in each of these diagrams one bulb is drawn above the battery and one below, this need not be so. You will have found out that just moving components on the bench doesn't alter the current as long as you don't alter the ones you use or disconnect any of them. So we could also draw the circuit like this:

Fig. 19.24.

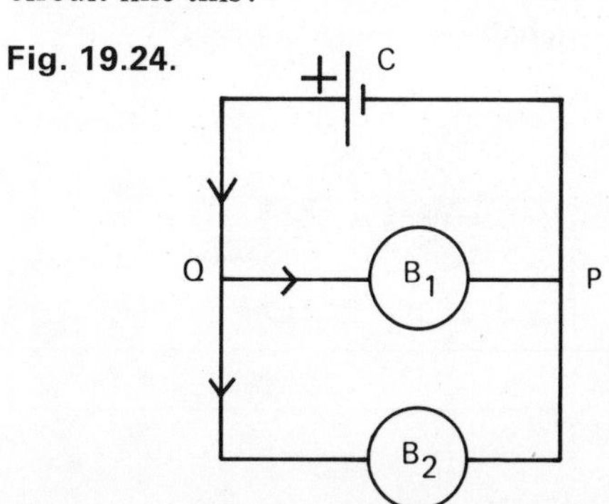

Whichever of these three diagrams you look at you can see that when the electrons get to the junction at P some of them go through B_1, and get back to the battery that way, while some go through B_2 instead. (Remember that the electrons go the opposite way to the arrows.) Those that go through B_1 don't go through B_2, and those that go through B_2 don't go through B_1. When two components are arranged so that the current can pass through either of them but not through both we say that the components are *in parallel.*

Experiments: *Components in parallel*

Wire up once again the circuit shown in Figs. 19.23 and 19.24. We know that the current that goes through B_1 doesn't go through B_2 and it would be interesting to find out how much current goes through each bulb. Suppose that the bulbs are the same, would you expect the currents to be the same? Would there be any reason why one bulb should get more than the other? We can check this using an ammeter. To measure a current with an ammeter you must make the current you want to measure run through it, so we shall need to break the circuit and then join it up again with the ammeter in between.

To measure *all* the current that leaves the cell C, where will you put the ammeter? – between P and C, Q and C, P and B_1, Q and B_2? – have you any other ideas? When you have decided, insert the ammeter and read the current.

Next think where you should put the ammeter to read the current through B_1. (There are two possible correct places.) Record this current. Finally record the current through B_2. Leaving the meter in the position to read the current through B_2, disconnect B_1. Does the current through B_2 alter at all? a little? a lot? Does the cell supply more electrons every second if both bulbs are there?

What do you suppose would happen in the end if you just let the bulbs go on burning? Would the chemical energy in the cell eventually all be used up? Would this happen more quickly if you let the cell light just one bulb? Would it happen more quickly if the cell lit two bulbs arranged in series?

Fig. 19.25.

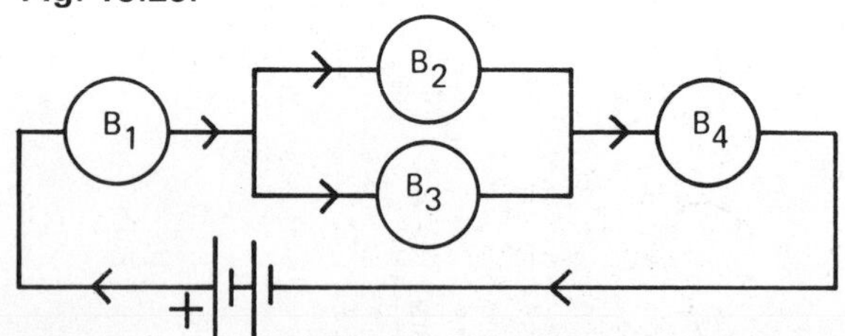

Here is another circuit you can set up. Look at the diagram carefully before you make the connections. Are the bulbs B_1 and B_2 in series? Are B_2 and B_3? Are B_3 and B_4? Are B_1 and B_4? Using an ammeter, find out how much current the cell is giving. Through which bulbs does *all* this current pass? Measure the current through B_2, through B_3. (Think carefully where to put the ammeter.)

If you have done all these experiments carefully you have learned many facts about the way in which cells push currents through circuits. If you connect one cell to two bulbs in series it can't push electrons through them quickly enough to light them properly. There are fewer ampères than with just one bulb. The bulb filament is made of very fine wire, so it is not easy to push electrons through it and if they all have to go through both bulbs it is naturally more difficult. We say that the bulb offers resistance to the passage of current, or simply that it has *resistance*. Two bulbs in series have more resistance than one bulb. If we put a second cell in series with the first to help push, we can make both bulbs light normally. But we must be careful to join the cells the right way round.

If we put two bulbs in parallel then there are two ways the electrons can go. It is like running water out of a bath through two plug-holes – it will run away more easily. So one cell can push more current through two bulbs in parallel than it can through one. Two bulbs in parallel have *less* resistance than either one of them. At first this seems strange, but if you think about it, it makes sense!

Experiment: *Can we have more than two components in parallel?*

(1) Draw again the circuit of Fig. 19.24. How will you connect another bulb B in parallel with B_1 and B_2? Will the cell have to supply more current? Will the resistance of the three bulbs in parallel be more or less than that of the two bulbs? Will it be more or less than that of two bulbs in series? than that of one bulb? Measure the current going through the cell with an ammeter and see if you are right.

How many cells would you need to light normally three bulbs arranged in series? How would you arrange the cells? Draw a diagram, then set up the circuit and see if you are right. Measure the current that is passing through the three bulbs in series.

(2) Study carefully the circuit shown in Fig. 19.8 (*c*). Put a switch into the circuit so that it will switch all

the bulbs on and off together. Can you put in a switch that will switch one bulb off and leave the others burning?

Next arrange the circuit of Fig. 19.23. Put in a switch that will switch off one bulb only – say B_1. Can you now put the switch somewhere where it will switch both bulbs off together? If you think so, try this and see if it works.

Questions

(1) Your classroom probably has several electric light bulbs in it – if not I expect you can think of a room that has. Can you switch on more than one at a time? Does this tell you whether they are in series or in parallel? If one bulb is taken out of its socket, can you still light up the other bulbs? Can you now say whether they are in series or in parallel? Will you use more current or less when you switch on more lights?

(2) A car has many lights and sometimes other gadgets such as windscreen wipers on it. Are they arranged in series or in parallel? Give your reasons.

Fig. 19.26. How many lights can you count on the front of this car? Are they arranged in series or in parallel? Name some other items of a car's electrical equipment and say how they are connected to the battery.

(3) You may have used fairy lights to decorate your Christmas tree at home. Are they arranged in series

Fig. 19.27. A peep inside the Concorde. Some of the 200 km or more of wire that are used to carry energy from the batteries and dynamos of the aircraft to the places where the energy is needed. Where does the aircraft get the energy that lifts it into the air and makes it travel faster than sound?

or in parallel? How can you test?

(4) A wise guy once thought 'They tell me that bulbs in series use less current than the same number of bulbs in parallel, so I will rewire all my bulbs in series and save money'. Why would he be disappointed if he tried?

Experiment

Your earlier experiments have proved that if you have one cell it can't push as much current through two bulbs in series as it can send through one, and with three bulbs it sends even less, and we decided why this was. It isn't only bulbs that have resistance, and here is a simple experiment to show this.

You will be given a length of fine wire and two drawing pins (or something of the kind). Wrap the wire round the spike of the drawing pin two or three times and push the pin into a piece of wood so that you can fix a croc clip to it. Do the same at the other end of the wire. You now have a length of wire XY (Fig. 19.28) stretched out along the bench.

Fig. 19.28.

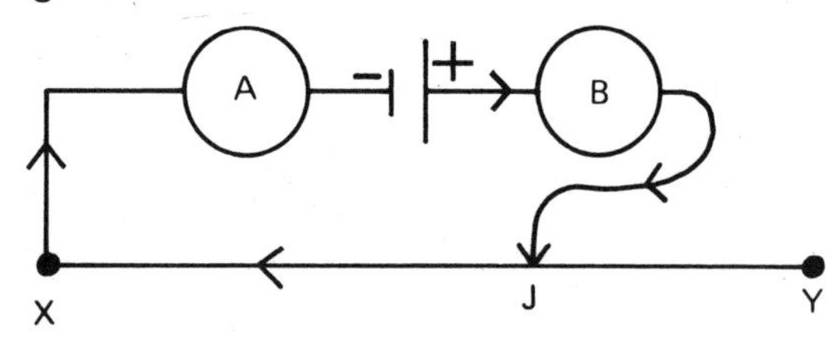

Complete the circuit as shown in the figure. J is a croc clip – or just a piece of wire – with which you can touch XY where you like, and the sign ↓ means this.

First touch the wire XY with J at Y. Does B light up brightly? dimly? normally? or not at all? Through what conductors is the cell pushing electrons? If you move J towards X, would the cell's job be easier, or harder, or the same as before? Move J in this way and notice what happens – look at the bulb and the ammeter. Where will you put J to get as big a current as possible? What differences would you notice if you had used a wire XY twice as long?

Fig. 19.29. Regulating the electric current. Where must the boy put the clip in his right hand if he wants the bulb to light up brightly?

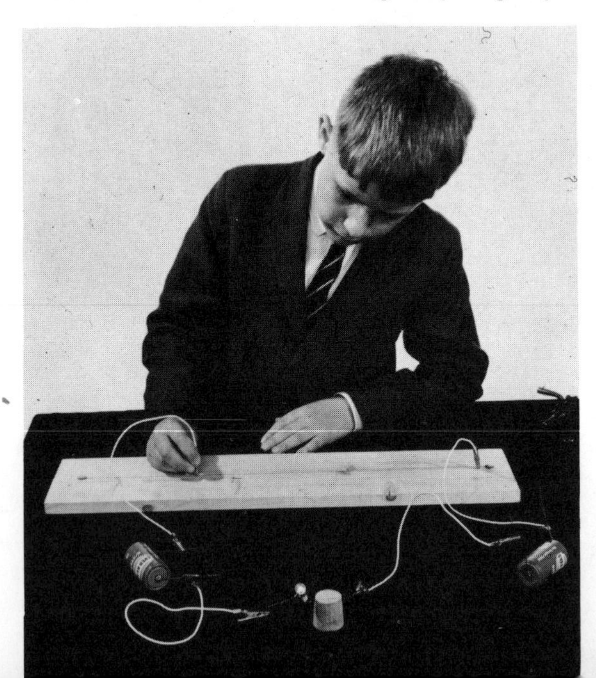

This experiment shows us one way in which we can vary a current gradually without having to put extra cells in or take cells out. The more wire we put in the circuit, the bigger the resistance is, and so the smaller the current. Our arrangement was a home-made one – electricians use a component called a *rheostat*. There are various kinds of rheostat. Here are two (Fig. 19.30).

Fig. 19.30.

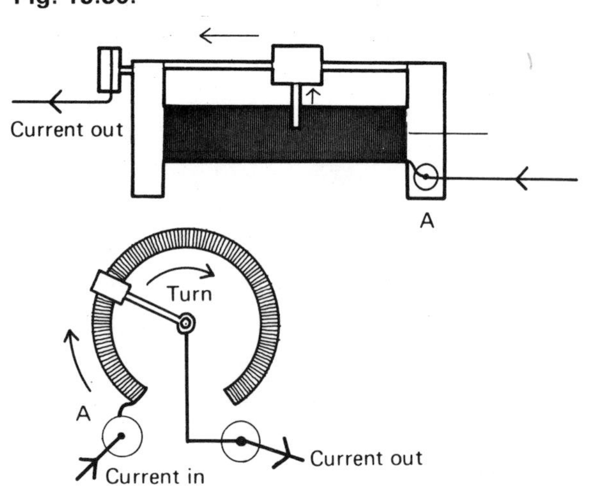

In each kind the current has to go through a certain amount of wire on its way through the rheostat. The wire is coiled instead of being straight like XY, and the slider takes the place of the croc clip.

When will the current be biggest – when the slider is near A or furthest away from A? Does it matter at which terminal the current goes in?

Experiment

When a bulb is lit by one cell it lights normally, and when two cells are used in series it lights more brightly. What will happen if we use three or four cells to light one bulb? Arrange a circuit like the one

in Fig. 19.28, but put three cells in series with a switch, one bulb and the wire XY. As before, start with J near Y, move it gradually towards X and watch carefully what happens – keep an eye on the ammeter reading. You would expect the bulb to light more brightly. If this is all that happens, try a fourth cell. If the bulb does not light after the experiment, examine the filament with a hand lens. What has happened to it?

This last experiment shows us that if we try to send too much current through a component we may damage it. You have seen that the more current you send through a flash-lamp bulb, the hotter the filament gets. A lamp filament is meant to get hot – it is made of an element called tungsten, and we said earlier (p. 30) that this is very hard to melt – but if we make the current too big then it *will* melt. The wires that carry current from place to place inside our houses are *not* meant to get very hot, and if too much current passes through them and they get too hot then they could set fire to the building. Electricians have to do something about this, and they use a little safety device called a *fuse*. You can do a simple experiment to see how this works.

Experiment: *How a fuse works*
Arrange a circuit like that of Fig. 19.28, but in series with the bulb put a length of the fuse wire given to you. Starting with J at Y, move it slowly as before, towards X, and watch the fuse wire. Does the bulb go out this time? Why? Test it with one cell to see if it is still in order. Has the fuse protected the bulb?

Repeat the experiment, but use one cell only and leave out the bulb. Watch the ammeter carefully as you move J. What is the biggest current the fuse can take? What is the normal current taken by one bulb from one cell? from two cells?

Would the bulb be as well protected if you used a thicker piece of wire? a thinner piece? If you use too thin a piece, will your bulb light up normally?

Fig. 19.31.

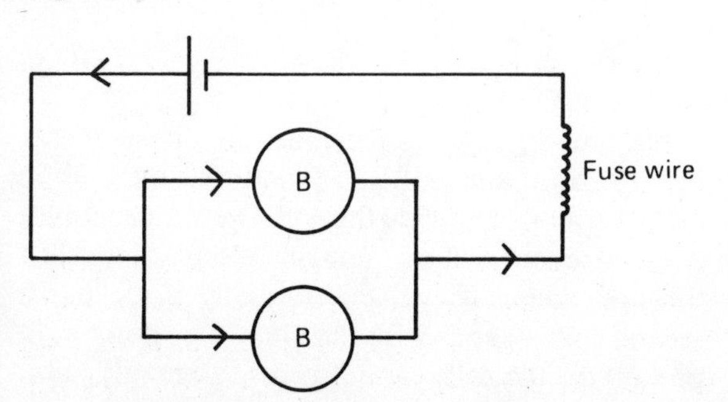

Suppose you used another piece of the same fuse wire in the circuit of Fig. 19.31, would the bulbs be properly protected? Would they light up normally? If not, would you use a thicker or a thinner piece of wire for the fuse in this circuit?

When electricians put the wires into a building they arrange for each of the circuits to have a fuse somewhere in it. Sometimes the fuses are arranged in a special fuse-box. Figure 19.32 shows a fuse of this kind. The piece of fuse wire can be seen. If it melts the circuit is broken and the current stops flowing.

Fig. 19.32.

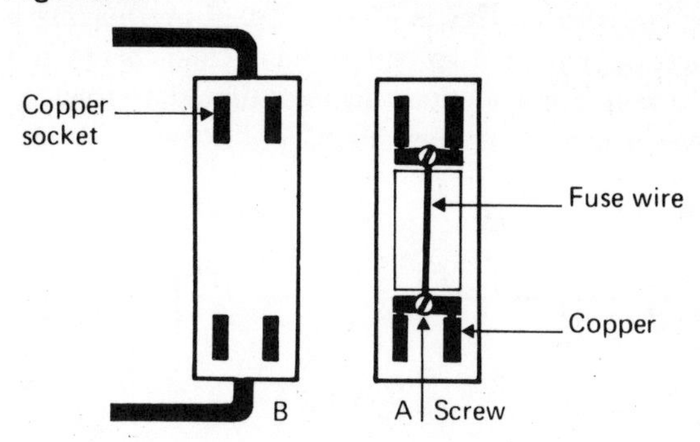

Another kind of fuse is shown in Fig. 19.33. The wire is fixed in a little tube, sometimes a glass tube, and the tube clips into the plug as shown. If this fuse melts you throw it away and put in another.

Fig. 19.33.

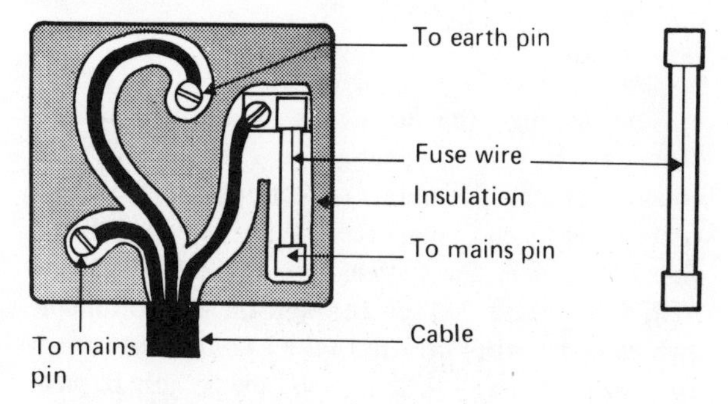

In each type you must choose wires of the correct thickness for the job – a 5 A fuse will melt if a current of 5 ampères flows through it, a 250 mA fuse melts if 250 milliampères flows and so on.
The symbol for a fuse is ⦵, or ⦵.

What makes the fuse melt?
We know that the fuse melts, or blows, because too much current flows through it, but why should this happen? In our experiment it happened because we

were using three cells to light one bulb, which wasn't really fair. But in houses, if we buy the right equipment, the current shouldn't get too big, and yet fuses sometimes blow. Experiments can tell us something about this, too.

Experiment

Arrange the circuit as shown in Fig. 19.34 (*a*), so that the cell sends current through B_1. The switch is not essential – you can use your croc clips for switching off. Now switch off, arrange B_2 in parallel with B_1, and switch on again. Next put B_3 in parallel with the other two, and if you like, more bulbs in parallel. What happens to the fuse if you make the number of bulbs too big? This is what is called overloading a circuit. If you switch on too many components in a room it can sometimes happen that the circuit is overloaded and the fuse blows.

Fig. 19.34.

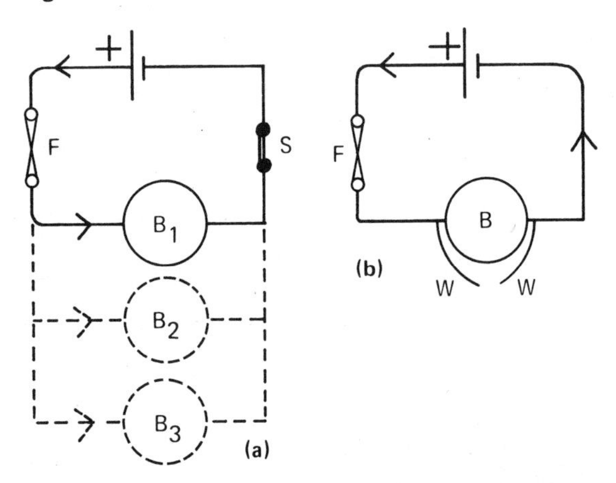

Now arrange the circuit as in Fig. 19.34 (*b*). W, W are two pieces of copper wire, one end of each touches one of the bulb connections. Now let the other ends touch one another. What happens to the fuse? Why does the current flow through the fuse? Will the current now go through the bulb filament (which is fine wire) or will it take a short cut through the thick wires W? Will the cell now be able to push more current round the circuit? Does this explain what happened to the fuse?

When this happens we say that the bulb has been *short circuited*, or just *shorted*. This can happen in a mains circuit if wires are carelessly screwed onto terminals and become loose and touch. It can also happen if the insulation on wires gets old or worn, or if the insulation gets heated too much because of overloading. The wires inside the cable may touch and the current will then go across the point where they touch instead of through the fire or cooker it is meant for.

If you put another fuse in place of the first without finding out what is wrong, then of course the second fuse will blow too. *You must switch off the current at the mains*, and then look for the trouble and put it right. You must not put a piece of ordinary wire instead of fuse wire in the fuse-box – if you do, the big current may go on flowing and set fire to some part of your home.

Summary

We can arrange electrical components in two ways – so that the same electrons go through both (series), or so that electrons go through one or the other but not through both (parallel).

Putting two components in series increases resistance, putting two components in parallel reduces resistance.

House and car lights are usually arranged in parallel, so that if one lamps fails the others go on burning.

A fuse is made of wire that melts easily and is put in a circuit to stop overheating if the current gets too big. You must use the right fuse for the circuit. When a fuse blows it is because too much current flows, due either to overloading or to a short circuit.

It is no good replacing the fuse unless you have found out the cause of the trouble. It must *never* be replaced with ordinary wire.

A fuse in a mains fuse-box should *never* be replaced unless the current has been switched off. (It is a good idea to keep a candle and matches near your fuse box, where you can find them in the dark.)

How hard does a battery push?

Fig. 19.35.

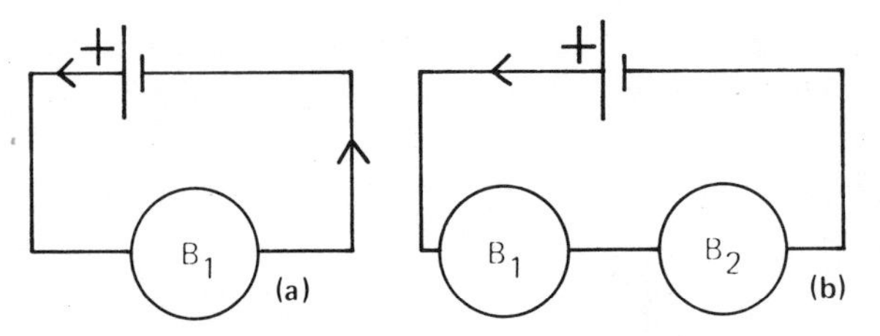

Some time ago you set up the circuits shown in Fig. 19.35 (*a*) and (*b*). You found that the bulb in (*a*) lit normally but in (*b*) the bulbs were dim because they were not getting enough electrons pushed through them. They light normally if we put a second cell in series with the first. They don't light properly if the cells cannot provide enough e.m.f.

(what does that mean?) to drive the necessary current through them. In fact, if the cell in (*a*) had been in use for a long time it wouldn't light the bulb satisfactorily because it would already have used up most of its chemical energy. It would be a pity to throw away a cell as useless unless we had made sure it was faulty, so we need an instrument to measure the pushing power of cells. An instrument for this is called a *voltmeter*, and it gives readings in units called *volts*. You need not worry just now how it works or how big a volt is. The symbol for a voltmeter is –(V)–

Experiment

You will need three or four dry cells. First arrange a circuit as in Fig. 19.36 (*a*), taking care to connect the + ve on your cell to the + ve, brown or red terminal, on your voltmeter. Record the voltmeter reading. Do the same for each of the other cells, making a table showing how many volts each cell gives. Did each of your cells give the same reading? Suggest three reasons why they might be different. If one of them gave a much lower reading than the others, which reason is most likely?

Fig. 19.36.

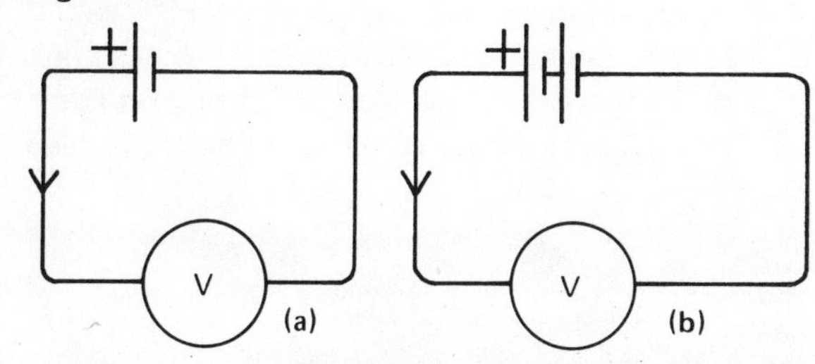

Now put two cells in series as in Fig. 19.36 (*b*), and record the volts given. Do this for two or three pairs of cells. Is the number of volts given by two cells equal to the sum of their separate readings? exactly? nearly?

Try this experiment again using three cells in series, or more if you are told that it is safe with your voltmeter. Do the readings 'add up'?

Work out from your observations the average voltage of a single cell.

(2) Arrange the circuit as shown. When the voltmeter is placed like this, it will measure the push between opposite sides of the bulb. Move J gradually from Y towards X. What do you notice about the bulb? Do this again but this time watch the voltmeter. How do its readings change as J moves towards X?

You know that when the bulb becomes brighter it

Fig. 19.37.

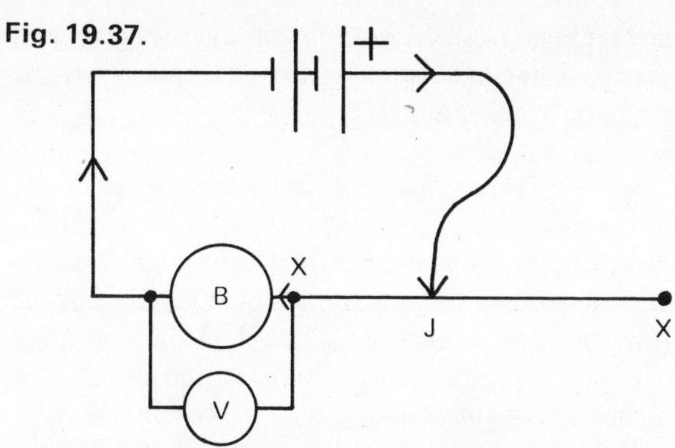

is because more heat energy is being produced there. So the voltmeter tells us when more energy is being used between the two points to which we connect it. Even when the filament is not hot enough to give out any light the voltmeter shows that some energy is being used there to warm it.

Experiment

Arrange a circuit with three cells in series and three bulbs in series. Will the bulbs' light be normal, bright or dim? Record the volts across each of the three bulbs and then across all three.

Does the same current pass through all three bulbs? Does it take roughly twice as many volts to push this current through two bulbs as it does through one? Are there roughly three times as many volts across three bulbs as across one? Does it take roughly the same amount of energy to heat each bulb? What connection is there between the number of volts between two points and the amount of energy used between them in this circuit?

Fig. 19.38.

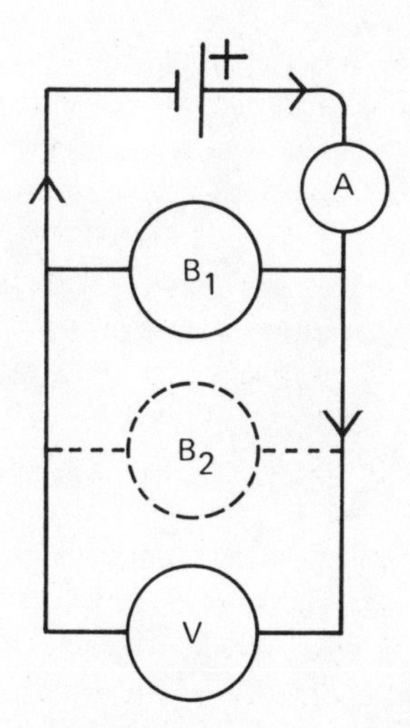

Arrange the circuit shown, and read the voltmeter. Put a second bulb in parallel with the first and read the voltmeter again. Is the reading now double what

it was? Has it altered much? Is the energy being used the same for two bulbs as for one? How has the ammeter reading altered?

These experiments tell us that we can't work out how much energy is being produced between two points in a circuit using only a voltmeter. We also need to know how much current is flowing. If we have two components in series then the voltmeter can tell us which of them is taking more energy, but if they are in parallel, it can't.

You may be interested to know that if a voltmeter connected between two points reads one volt, then for every ampère that flows one joule of electrical energy is being changed into heat (or some other kind of energy) every second, in between the points.

We have sometimes used the word cell and sometimes battery. If you want to be quite correct you should use battery only when you have more than one cell. However more people talk about dry batteries than about dry cells, so that it is not very important. You may be interested to know what is inside the dry cells we have been using. Here is a diagram of one (Fig. 19.39).

A new cell of this kind gives about 1·5 volts. If you can get hold of an old discarded cell it is interesting to cut it in two with a hacksaw and you can then see for yourself what is inside it.

Fig. 19.39.

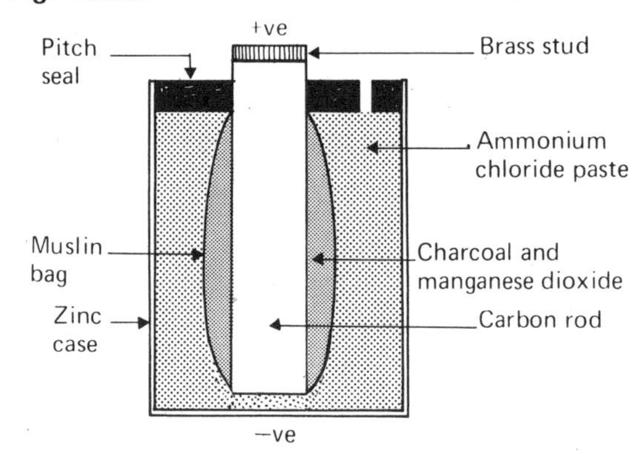

Car batteries

If your family have a car, ask your father to let you look at the battery, which is quite different from the one in Fig. 19.39, and is shown in Fig. 19.40. The battery is made from six cells arranged in series. When the engine is running at normal speed some of its energy is used to make electrical energy in a dynamo, or generator, rather like the one we used on p. 102, only bigger. Some of this energy is used to light the lamps and so on, but some of it goes into the battery, where it is changed into chemical energy. When the car stops, the chemical energy in the battery changes back into electrical energy and keeps the lamps lit when the dynamo isn't working. A battery of this kind is often called an *accumulator*, because energy is stored up or accumulated in it so

Fig. 19.40. Changing energy under the bonnet. The fan belt F transfers some of the motion energy of the engine to the dynamo D, where it is changed into electrical energy. Some of this goes to the accumulator A where it changes into chemical energy and is stored until it is needed.

that it can be used later if needed. Car accumulators are usually of the lead acid kind, but there are other kinds, and you should never tamper with a car battery or touch it without permission. Each cell of a car battery gives about 2 volts. How many volts does the whole battery give?

A model fire alarm

Fig. 19.41.

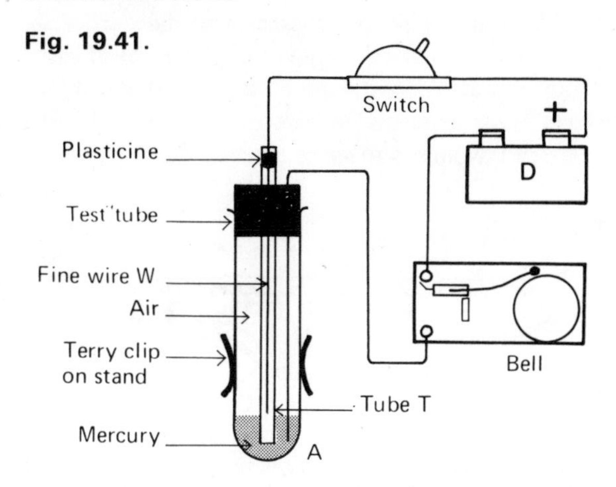

A simple model fire alarm that you can make yourself is shown in Fig. 19.41. The test tube A is fitted with a rubber bung through which pass a wire and a piece of narrow glass tubing called capillary tubing. Inside the capillary tubing is another wire that doesn't reach as far as the mercury. D is a dry battery. Instead of the bell you can use a flash-lamp bulb with red paper round it to show a warning light.

Why doesn't the bell ring when you switch on? If the test tube gets hot, what happens to the air in it? What does the air do to the mercury? Why must the bung be a tight fit? (It is a good idea to grease the bung). When will the bell start to ring? Why is it best to use a narrow tube T? What will happen if W is moved higher up the tube?

When you have completed the model, wire up the circuit and warm the test tube by standing a beaker of hot water near it. You will then see if you correctly worked out what will happen.

More experiments

(1) You may have heard something described as silver plated or chromium plated – the bumper bars on motor-cars are often chromium plated. They are made of iron or steel but they have a thin layer of the element chromium on the outside to stop them rusting. When iron rusts it combines with – what element in the air? Chromium doesn't easily combine with this element so it doesn't rust. Most plated things are electroplated, which means that the thin layer on the outside is put on by a chemical change caused by electrical energy. You may like to do some copper plating.

Half fill a jam jar with distilled water (you can

Fig. 19.42. To make iron shiny and to stop it from rusting we sometimes plate it with a layer of chromium. Which parts of a car are chromium plated? Chromium is an element and we get it by breaking down certain chemicals using an electric current. The lady at the back is arranging things in a vat where the plating happens.

Fig. 19.43.

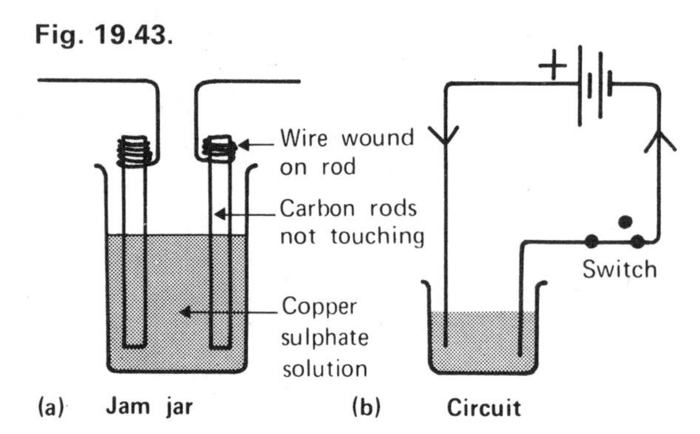

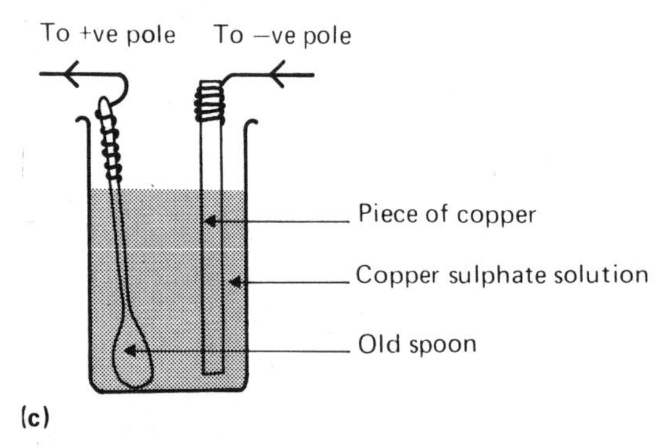

get some from the bottom of the fridge) and in it put enough crystals of copper sulphate so that a saturated solution is made. (What does saturated mean?) Add a little dilute sulphuric acid. While the crystals are dissolving take two carbon rods (you can get them from old flash-lamp batteries), and wind a piece of copper wire round one end of each as shown in Fig. 19.43, dip these in the copper sulphate solution but do not let them touch each other. Connect these wires to two or three cells arranged in series as in (*b*). Leave your circuit for about ten minutes and then break the circuit and examine the two carbon rods.

Is either rod changed in any way? If so, in what way? Is the other rod changed? Did you notice anything else happening as the current passed? Make a note of your observations like this:

	Appearance	Has it changed?
Rod connected to +ve Rod connected to −ve		

When you have examined these rods carefully, replace them in the solution as before but connect up so that the −ve pole of your battery is now connected to the rod that was +ve before. Pass current again for about 15 min. and examine the two rods again. What has happened to them now?

If you wish to copper plate an *old* spoon or fork (make sure that it is one for which your mother has no further use) use the same solution of copper sulphate but instead of the rods connect a piece of copper to the +ve pole of the battery and the old spoon to the −ve pole and dip them both in the solution while the current passes. Fig. 19.43 (*c*).

(2) Make a simple compass as you did earlier by pushing a magnetised needle through a piece of cork. Take a saucer or a shallow dish, wind twenty or thirty turns of insulated wire round it (Fig. 19.44). Leave a few inches to spare on each side.

Fig. 19.44.

Fig. 19.45.

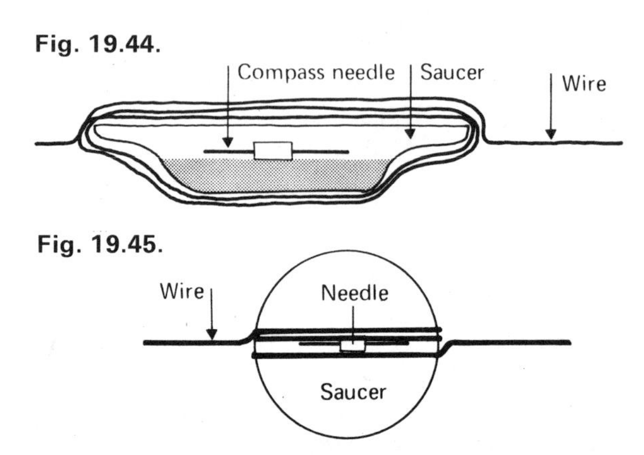

Put in the saucer enough water to float your compass needle and allow it to settle. Turn the saucer very carefully until the wires are in the same direction as the needle (Fig. 19.45). Connect the spare wires to a battery – one cell should do. What happens to the compass? Does it move? Try again with the battery poles reversed. What happens this time? Do you think that an electric current has some magnetic properties? What kind of energy does the current give to the compass when it is moving? when it is still?

Fig. 19.46.

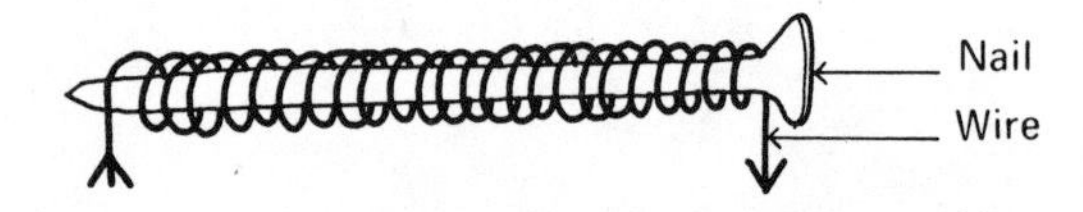

(3) Wind twenty or thirty turns of insulated wire round a big iron nail or an old iron bolt, or other piece of iron. It doesn't matter if the iron is rusty. Test the iron with paper clips or tintacks to see if it is magnetised. Connect up your dry battery and test the iron while the current is passing round it. Has the iron changed? Switch off the current and test again. Is the change in the iron physical or

chemical? (You must not use an accumulator for this experiment).

The experiments show that the energy carried by an electric current does not always appear as heat energy.

(4) Figure 19.47 shows a very simple kind of electric motor called Barlow's wheel. A little mercury is placed in the cut in the wood and the spikes on the wheel dip into this. Two magnets are fixed with opposite poles on each side of the slit. One wire goes to the wheel, the other to the mercury in the cut.

What happens when the current is switched on? What kind of energy does the current give the wheel? What happens if you reverse the current? What happens if you change round the poles of the magnets?

Fig. 19.47. Barlow's wheel. The wheel is made of copper. The mercury is put in the small plastic box that fits the slot in the wood. You can see one of the magnet's poles – the other is hidden behind the wheel.

Summary

An electric current carries energy from one place to another.

A battery changes chemical energy into electrical energy, a dynamo changes motion energy into electrical energy, a motor changes electrical energy into motion energy.

The energy carried by a current may change into chemical energy (as in electroplating) or into heat (as in a fire) or into motion energy (as in an electric motor).

An ammeter is an instrument for measuring how fast electrons are going round a circuit. A voltmeter measures how hard electrons are being pushed through a component.

Chapter 20 **Oxygen**

We saw earlier that only a small part of the air takes part in burning, and this is called oxygen. Oxygen is such an important and useful chemical that we must learn more about it. You may already know some uses for it and you could look more up in an encyclopaedia. Later on we will come back to these uses.

Here are some questions we must try to answer:
How much oxygen is in the air?
Can we take it from the air?
Can we make it from other chemicals? (You ought to know the answer to this already.)
What is it like when it is on its own? Or as scientists say, what are its properties?

How much oxygen is there in the air?
What fraction of the air was used up when we burned a candle in it? What fraction when phosphorus was burned in it? Did you get the same answer for both of them? Ought you to get the same? Were the experiments accurate? Did you measure carefully? We must now try to improve on this.

When you saw the bell jar – candle experiment you may have found it difficult to measure the fraction used up because the jar was a bad shape. We could measure the volumes more easily if we had a tube marked in cm^3 like a measuring cylinder or a burette. Which measures more accurately, a measuring cylinder or a burette? Why is it an advantage when the tube is narrow? In the experiment you will use a graduated tube (what does this mean?) marked up to $50 cm^3$.

Make a small loop at the end of a piece of copper wire (Fig. 20.1 (*a*)). Using a pair of tongs, take a small piece of phosphorus and put it in an evaporating dish of water. Remember that you must not touch it with your hands. Now wedge the phosphorus in the wire loop, Fig. 20.1 (*b*), and keep it under the water still. Next push the phosphorus up the graduated tube, (*c*), and put in a thin piece of rubber tube, (*d*). Stand the tube in water, (*e*), clamp it, and take away the rubber tube, (*f*). Read the level of water in the tube and write it down. Read the level the next day and again a day later. Go on doing this until the level stays the same. Why? Now you can work out the fraction of oxygen. What was the volume at first? What was the volume after? Find the volume used up and then its fraction of the whole. Set down the results:

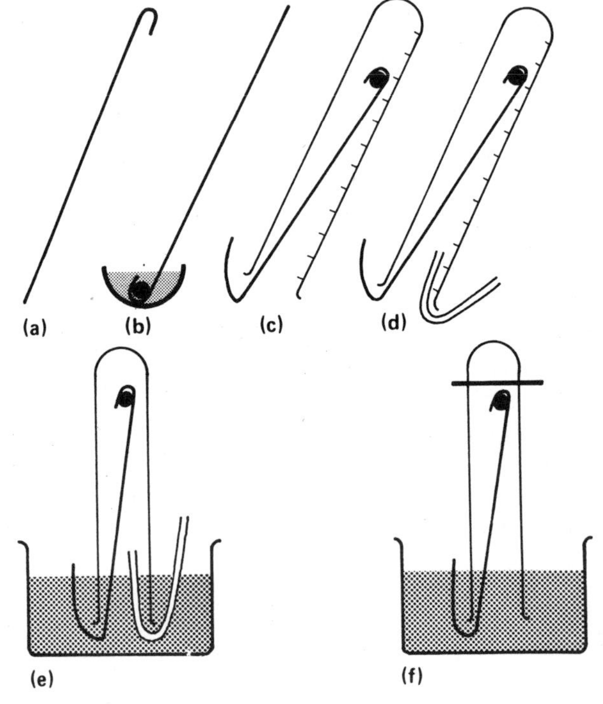

Fig. 20.1. Finding the fraction of oxygen in the air.

Volume of the air was cm³
Volume of the remainder was cm³
Volume of oxygen was cm³
Fraction was .

To be quite sure of our method we ought to put some phosphorus in a tube of oxygen. What will happen if our method is right? Do this experiment, Fig. 20.2.

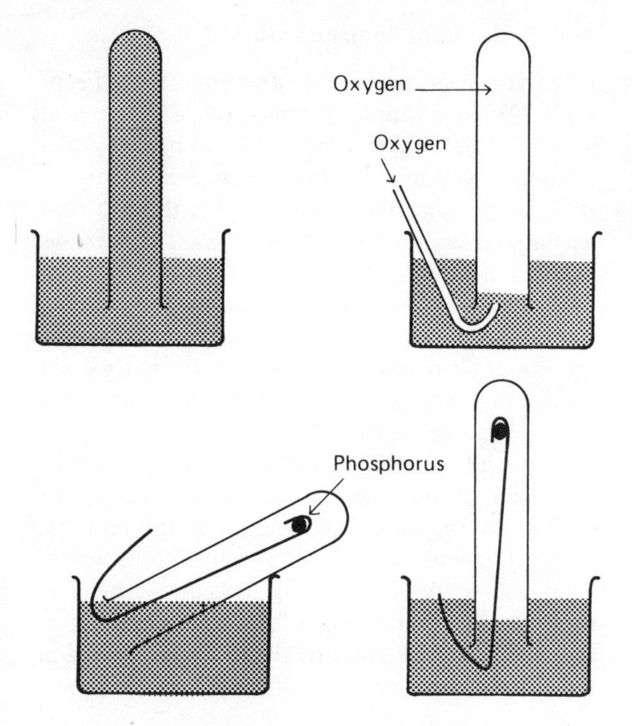

Fig. 20.2. Checking on our method.

Will the fraction of oxygen always be the same? What would happen if you closed all the windows in the laboratory and then lit all the Bunsen burners? The fraction of oxygen does change a little from place to place and from time to time, but it is roughly $\frac{1}{5}$. This is a useful figure to remember.

Another way of finding the fraction of oxygen in the air.

In this experiment you will use a chemical called pyrogallic acid. This can combine with oxygen very quickly, particularly if sodium hydroxide is mixed with it. You can see a picture of a pyrogallic acid molecule in Fig. 20.3. What kinds of atoms are in the molecule? How many of each are there? First we must check that the chemical does take oxygen from the air.

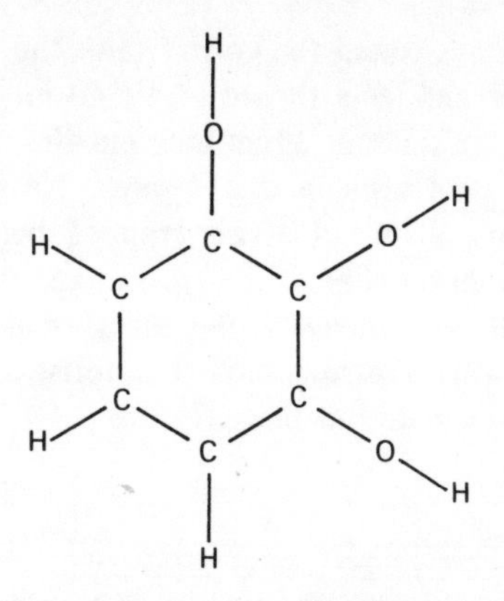

Fig. 20.3. The pyrogallic acid molecule.

Take a flask with a well-fitting rubber bung. Remove the bung and hold a lighted splint inside the flask. Does the splint go out immediately? Put one spatula of pyrogallic acid in the flask and then add $\frac{1}{4}$ test tube of sodium hydroxide solution. Put the bung back firmly and swirl the flask for two minutes. What do you see? Stand the flask on the bench and let the liquid drain to the bottom so that you can see into it. Now light the splint again and try it in the flask. Does the light go out immediately? We can now go on to use this chemical for measuring oxygen.

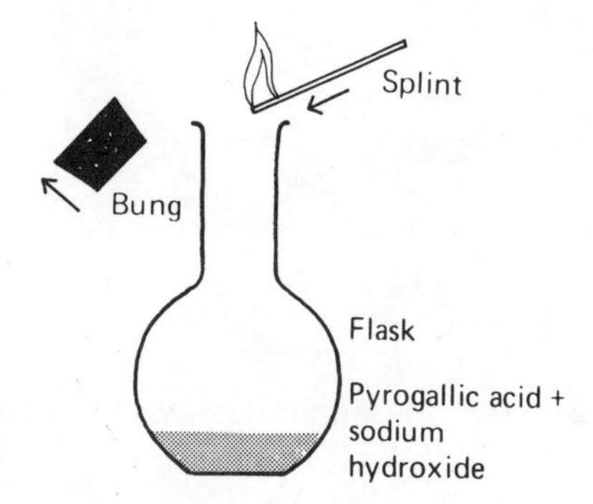

Fig. 20.4. Testing pyrogallic acid.

You will use a special tube which has a very fine hole through it – we call it a *capillary tube*. You may know that the fine blood vessels which join arteries to veins are called capillaries too. The tube is an angle with one side about 40 cm long, and the other 10 cm. See Fig. 20.5 (*a*). A rubber teat is fitted over the long side. You also need a rule and two small beakers, one filled with water and the other with pyrogallic acid in sodium hydroxide solution. (2 spatulas to 50 cm³.)

Put the short end into the water, squeeze the teat to force some air out and then relax the pressure so that water goes into the capillary. As the thread

of water goes round the corner take the tip out of the beaker and let a thread of air go in. When the thread of air is about 20 cm long put the tip back in the water and take in more water. Now the tube should have a thread of air trapped between two threads of water (Fig. 20.5 (*b*)). Put the tube on the metre rule and measure the air thread carefully (Fig. 20.5 (*c*)). You may have to practise a few times before you can do this properly.

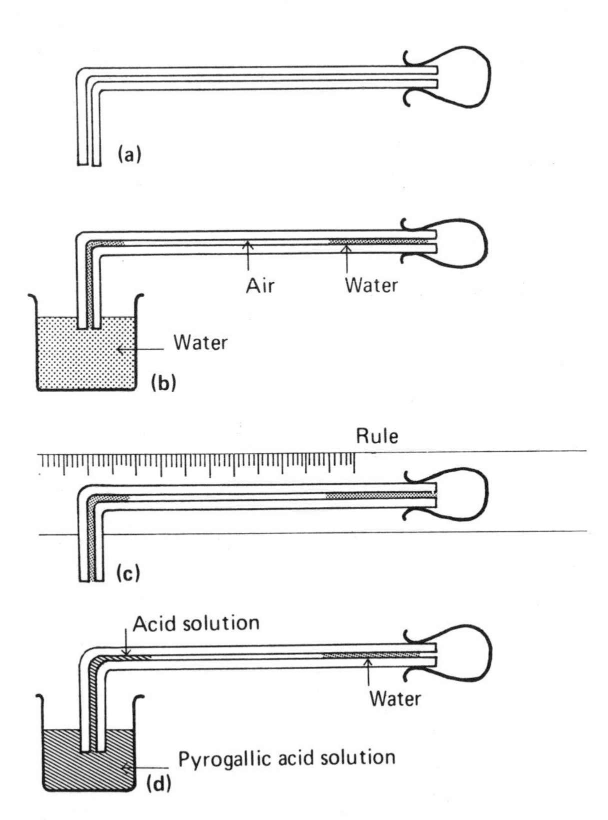

Fig. 20.5. The pyrogallol way of finding oxygen.

Now put the short end into the pyrogallic acid solution, squeeze the teat until the air is about to be pushed out – but not quite. Relax the teat so that the solution now goes into the tube (Fig. 20.5 (*d*)). Now squeeze and relax the teat so that the air thread moves to and fro over the 'pyrogallol wet glass'. Measure the bubble every few seconds until two lengths are the same (why?).

Set down your results:

Length of the air bubble was cm
Length after removing some oxygen was
After more……
Final length was cm
Length used up was cm
The fraction of oxygen is

Oxygen and how we keep warm

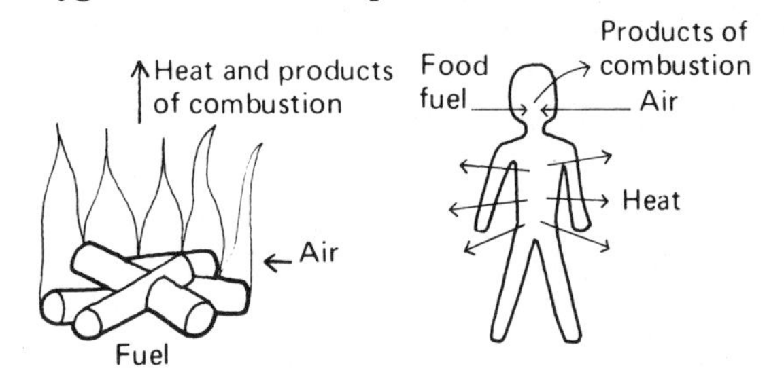

Fig. 20.6. What keeps us warm?

Have you ever wondered how your body keeps warm? What is your body temperature? If you heat up metal or stone or wood, they will begin to cool as soon as you take the heat away. But your body stays warm year after year. Why is this? Is heat leaving our bodies all the time? What are the ways in which heat will leave us? When you jump into a cold bed you soon warm it up. Heat energy is leaving our bodies all the time but our temperature stays the same. Of course, this must mean that we are *making* heat energy all the time. We also make other kinds of energy – name two kinds.

Chemical changes are happening in every part of our body all the time. Just as the slow rusting of iron in the thermos flask made it warm, so there are slow reactions inside us which keep us warm. Chemical energy is changing to heat energy. The food you eat is changed so that it can dissolve in the blood; the air you breathe in also dissolves in the blood. In every part of the body:

$$\text{food} + \text{oxygen} \xrightarrow{\text{gives}} \text{waste products} + \text{heat.}$$

Have you ever heard the saying: 'my ears are burning'? Well they really are, but very slowly!

Fig. 20.7.

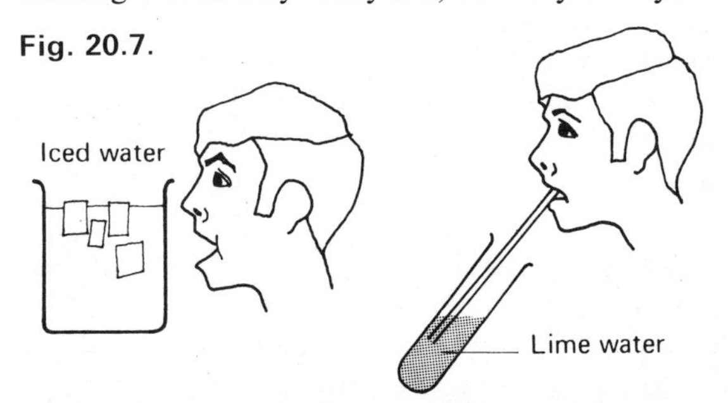

What products of combustion do we make?

Fill a dry beaker with iced water, then breathe on the outside of it. What do you see? When food burns in our body, some water is made.

Half fill a test tube with limewater and then blow through it using a drinking straw. What happens?

We also make a lot of carbon dioxide when food burns. You may be thinking that the carbon dioxide was in the air all the time and that we breathed it in and then out again. How can you check up on this? Do it.

Food and air react to give energy and carbon dioxide and water. You may remember that a candle behaves in the same way when it burns. But we do not give out light or soot! Nevertheless, the similarity is remarkable, isn't it?

Some uses of oxygen

Figure 20.8 shows it being used in a hospital. The boy is being anaesthetised before an operation. The anaesthetic gas (nitrous oxide) is mixed with oxygen before the patient breathes it. You can see the dials on the oxygen cylinders on the far side of the trolley. Why must you mix the anaesthetic with oxygen? Why is this better than with air?

When men go under the water or onto high mountain tops they take small cylinders of oxygen with them – sometimes the gas is air with more than the usual amount of oxygen in it.

This kind of breathing apparatus is used by firemen who must go into smoky buildings, or by rescue workers who go into mines full of poison gases. You can see how the cylinders fit on the worker's back. Notice the masks they wear. Fig. 20.9.

Welding is a way of joining metals together to

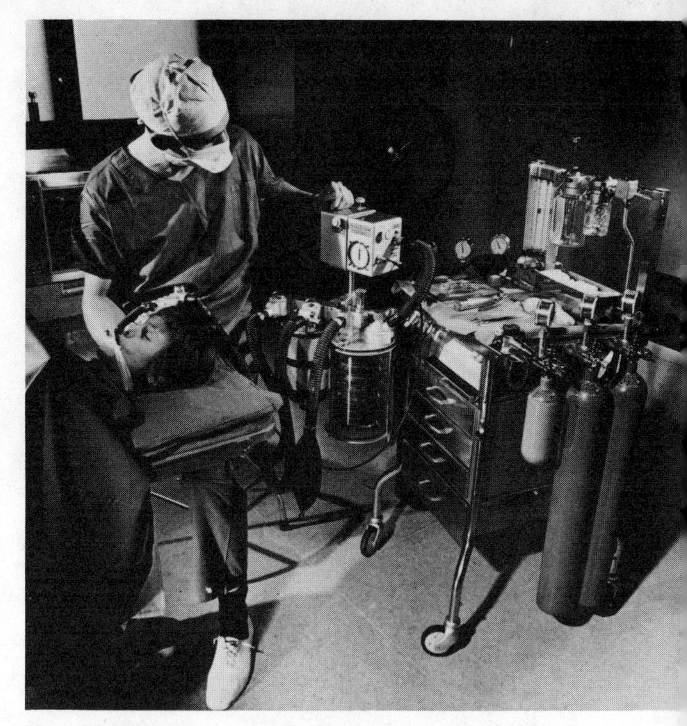

Fig. 20.8. Here a doctor is keeping a patient asleep. To do this he is giving a mixture of oxygen and anaesthetic. Why doesn't he give anaesthetic alone? Where are the gases? The apparatus is complicated because he must be able to control how much of each gas is being given. Why is the doctor wearing a mask? Why is the trolley made of metal?

Fig. 20.9. An instructor shows boys how rescue workers can breathe in coal mines which are full of smoke and poisonous gases. What are the valves under his left arm for? What is the hooter for? Will he need to carry a torch?

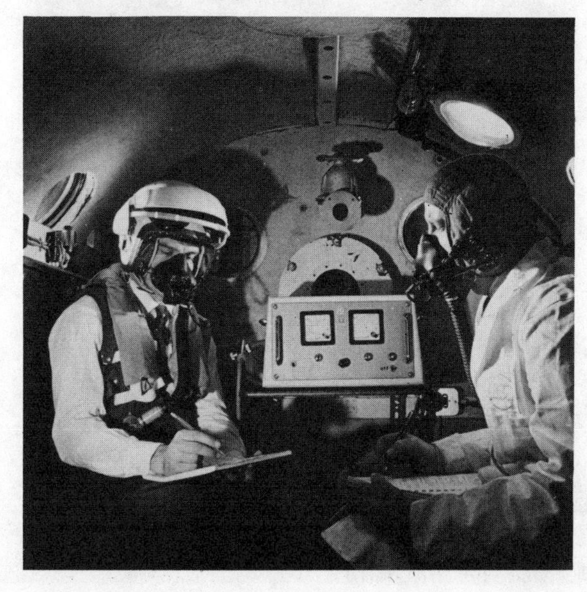

Fig. 20.10. These scientists are testing oxygen masks to make sure that they are suitable for airmen. The masks must be comfortable and they must have microphones built into them. In high flying airliners the passengers do not have to wear oxygen masks – why?

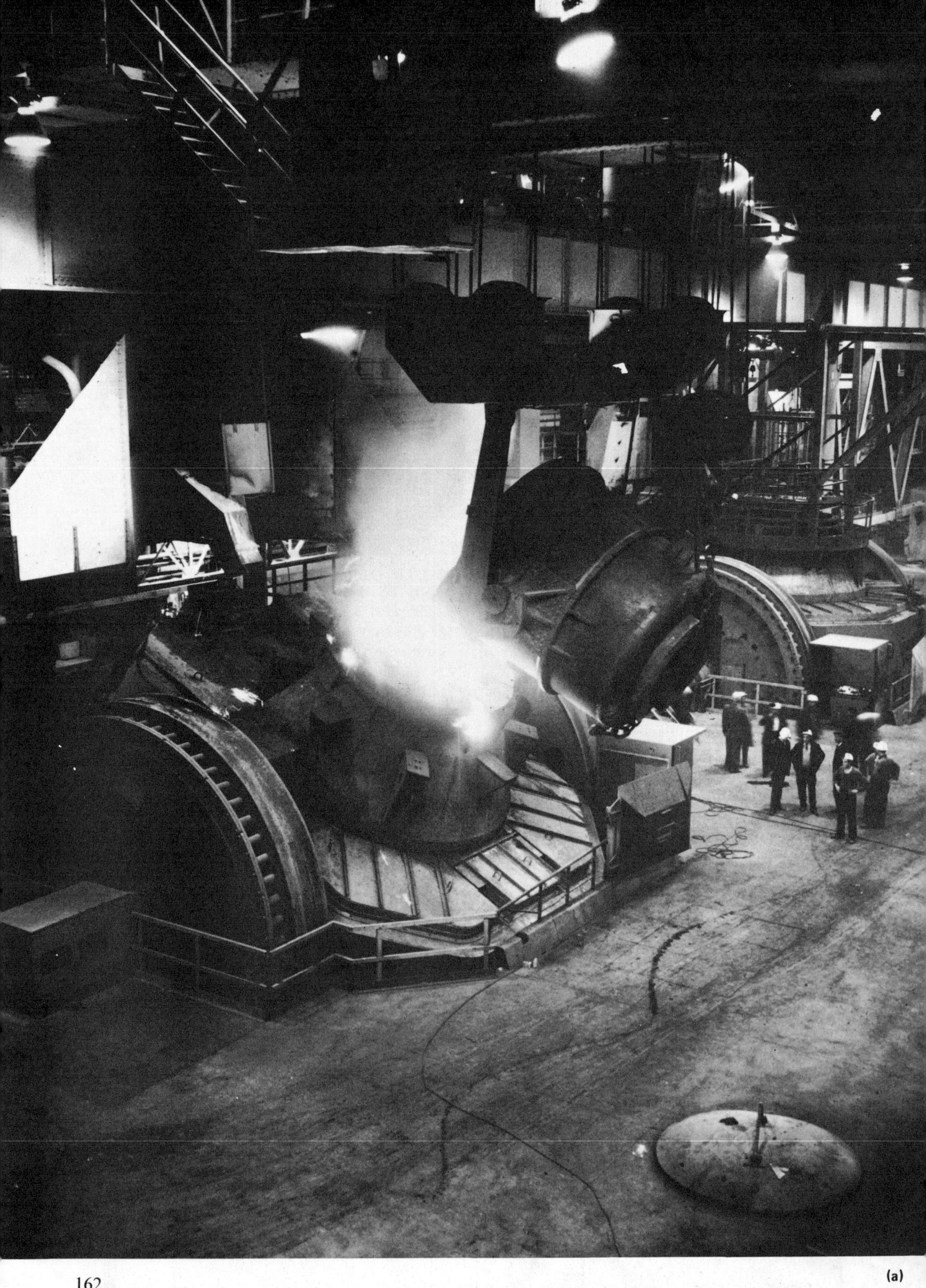

(a)

Fig. 20.11. An oxy-acetylene torch is being used to bore into the steel lining of a tunnel. Why is the workman wearing goggles?

make a very strong joint. In Fig. 20.11 you can see an engineer welding. A gas called acetylene is burning in oxygen and the flame is hot enough to melt iron so that the two pieces run into each other. Notice how the engineer protects his eyes from dazzle and sparks.

Figure 20.12 shows another use of the gas. Making iron and steel uses up a lot of oxygen. Iron ore is a rock which has a lot of iron in it. It is mixed with coke and limestone, and the three are heated up and then oxygen is blown through them – this makes them much hotter still. At this very high temperature the iron melts and runs to the bottom of the furnace to be collected. A furnace like this one will use many tonnes of oxygen every week. If a furnace uses 1 500 000 kilogrammes of oxygen, what will be the volume of the gas? (Density of oxygen is 1·5 grammes per litre.) What volume of air would do the same job?

Fig. 20.12. When iron is first made from the ore it contains a lot of impurities and these make it very brittle – so brittle that it will smash if it suffers a sharp blow. To make strong steel the impurities must be burned out of the iron and this is done in a converter. (*a*) Here you can see molten iron being poured into a converter ready for the start. How big is the 'bucket' pouring the iron? (look at the men). Can you see the man who is doing the pouring? (*b*) To make the impurities burn quickly a pipe is lowered into the mouth of the converter and oxygen is pumped through. The picture shows the brilliant flame and sparks as the burning takes place. You cannot see the oxygen pipe because of the glare. Why are the men wearing helmets? Can you see the instruments which tell the steelmakers what is happening inside the converter?

(b)

Fig. 20.13. To get oxygen from the air we must cool down the air so that it becomes a liquid. This is being done in the machine at the centre of the picture. It is white because of the frost on it. The scientist is looking at instruments which tell him the temperature and pressure of the air in the machine.

How we get oxygen from the air

You know that one-fifth of the air is oxygen. Getting it out is a most important industry because it is so useful. We do this by making the air into a liquid and this is done by cooling it down until it is very cold indeed. The liquid air we have now made has some liquid oxygen in it, but it also has some other liquid gases. Have you ever separated a mixture of two liquids? What does distillation mean? Liquid oxygen has a different boiling point from the other liquids in the mixture, so that when we distil the liquid air we can get the oxygen by itself and then put it in cylinders.

Something to do at home.

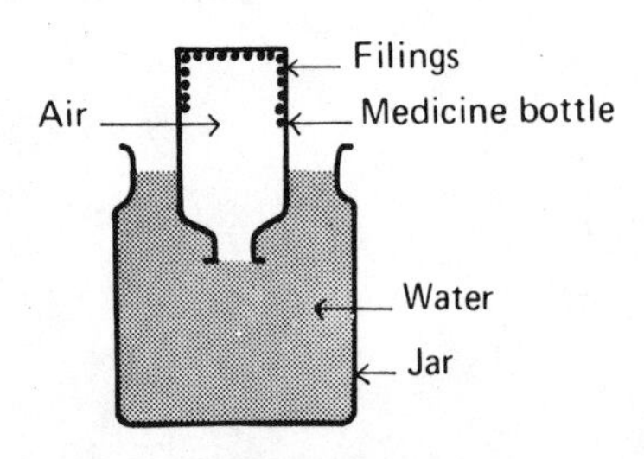

Fig. 20.14. The rusting of iron.

Clean out a medicine bottle and pour out the water. Then, while the inside is damp, sprinkle in a few iron filings so that they stick to the glass walls. Fill a jar with water and put the neck of the bottle into it (see Fig. 20.14). Hold the bottle like this as best you can–it will have to stay for a few days. The lid of the jar may be useful for making a wedge.

Why is *burning* like *rusting*? Why is *rusting* like the way in which our body uses food?

Questions

(1) What is made when the following elements are burned in air: magnesium, iron, copper, zinc, sodium, calcium? What is made when these metals rust?

(2) Give two reasons for thinking that air is a mixture and not a compound.

(3) Many years ago a scientist built the apparatus shown in Fig. 20.15. We would not do this experiment because the mouse would suffer a lot. Write down what you think would happen.

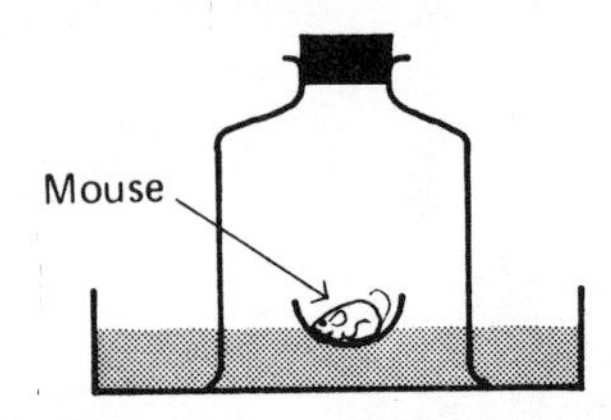

Fig. 20.15. The mouse experiment.

(4) Why does painting metal help to prevent it from rusting? Write down another way of stopping it from rusting.

(5) Explain carefully how you would find out whether a sample of oxygen had another gas mixed with it.

(6) The rusting of iron, the burning of wood, the way we use our food seem to be very different things. What is alike about them? How are they different from one another?

(7) We can stop iron from rusting by galvanising it or by chromium plating it. Use your library to find out what these words mean, and write about how we do them.

Summary
About one-fifth of the air is oxygen, and this is used up when things burn. Inside our bodies, food burns slowly to give us energy, and water and carbon dioxide are also formed. Rusting is a very slow process, but like burning, it uses up the oxygen of the air, and it makes heat energy. We get oxygen from the air by cooling it until it becomes a liquid, then distilling the liquid. Oxygen has many important uses in all kinds of industry.

Chapter 21 Other Gases in the Air

When things burn they use up a small part of the air. What do we call this part? The larger part did not react with burning things (what do we call a chemical which does not react?), and we must now find out more about this part. For instance, is it an element or a mixture or a compound?

When your mother hangs wet clothes on the line where does the water go to? When you burn gas or oil what are the products of combustion? What happens when you breathe out? All the time water and carbon dioxide are going into the air. Would you expect to find them in the inert part? If they are there will there always be the same amounts of them? Think about the weather.

Is water in the air?

If you have a salt pot you will know that the salt sometimes sticks in the hole at the top of it. This is because the salt sometimes gets damp. Why does it get damp? Where does the water come from? Table salt seems to be able to take water from the air. Are there other substances like this?

Your teacher will give you a list of chemicals. Make a label for each one which gives the name and date – and put your own name on it too. Cover the bottom of a small beaker with the first chemical and stand it on the label. Do this for the other chemicals and leave them all for a week.

Make a list of chemicals which get wet. Chemicals which take in water from the air and get wet are called *deliquescent* chemicals. Write a heading for your list 'deliquescent chemicals'. You have been told that when using chemicals you must always replace the lid immediately and now you know one reason for it. Can you think of any other reasons?

How much water is in the air today?

When you have a bath you use a towel to dry. Afterwards it feels wet of course. Does it weigh more? Why? Deliquescent chemicals take water from the air. Suppose we shut them in a closed space, how long could they go on taking in water? How could we tell how much they have taken in?

We need a big jar for this experiment and it is best if it can hold at least 5 litres (you may see them with distilled water in). When the jar is clean and dry you are ready to start.

Hang a test tube on a piece of thread and half fill the tube with calcium chloride. Now weigh the tube, thread and calcium chloride all together, using your top pan balance. Record the mass and the date.

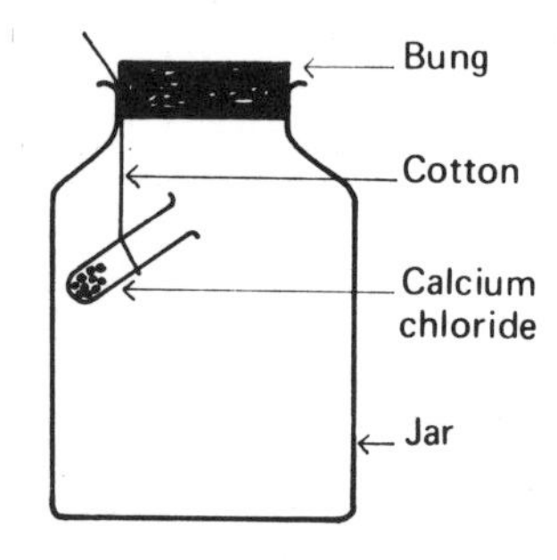

Fig. 21.1. How much water is in the air?

Lower the tube into the jar and then put in a bung very firmly so that it traps the thread and the tube hangs down (Fig. 21.1). After a week take the tube out and weigh it again.

Now you must find the volume of the air in the jar. If you are good at arithmetic you can work it out from the height and the diameter. But if you don't like this idea, find out by filling the jar using the biggest measuring cylinder that you can get. With a

two litre cylinder it will take about $2\frac{1}{2}$ lots. Make sure that you write down how you get your answer and what the answer is. Your results could be set out like this:

Mass of tube, cotton, calcium chloride

	before =	**g**
	after =	**g**
Volume of the jar	=	**cm³**

Now answer the following questions:
(1) What was the increase in mass?
(2) What was the mass of the water vapour?
(3) What was the mass of water vapour in each litre of air?
(4) Are you sure that the calcium chloride took in all the water? (What can you do about this?)
(5) Are you sure that the calcium chloride only takes in water? How would you check that it does not take oxygen? carbon dioxide?

More about molecules

Water is a compound and its molecules have two kinds of atoms – oxygen and hydrogen. You can see a picture of a molecule below.

H—O—H

How many atoms are there altogether? These molecules move about in the air, and of course they are much too small to see. But as the air cools down the molecules move more slowly. They gather in large numbers and we see drops of water (a mist).

Fig. 21.2. Sometimes the water in the air forms a mist; tiny drops of water hang in the air and prevent us from seeing clearly. In towns the droplets gather soot and this makes a fog. Here is a fog in a station in London. Why do smoky chimneys make fogs? Write down two good reasons why we should use smokeless fuels.

Finding carbon dioxide in the air

What is the test for carbon dioxide? See p. 77. If it is in the air and we bubble air through limewater it ought to go cloudy so we will do this to see if we are right. We may have to bubble a lot of air so we will use a pump to do the job for us. Why mustn't we blow through it? You can see a diagram of a water-powered pump in Fig. 21.3. Water rushes through the pump and pulls air with it – just like a big fast lorry pulls air after it. Have you ever seen paper on the road following a lorry?

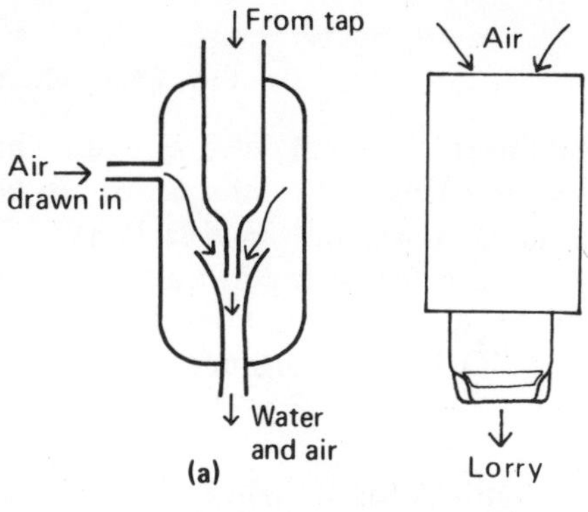

Fig. 21.3. A water pump.

Put some clear limewater into a wash bottle, and connect this to the suction pump (Fig. 21.4). Turn on the water to make the pump work and get a slow stream of air bubbling through the limewater. Leave it working for a long time.

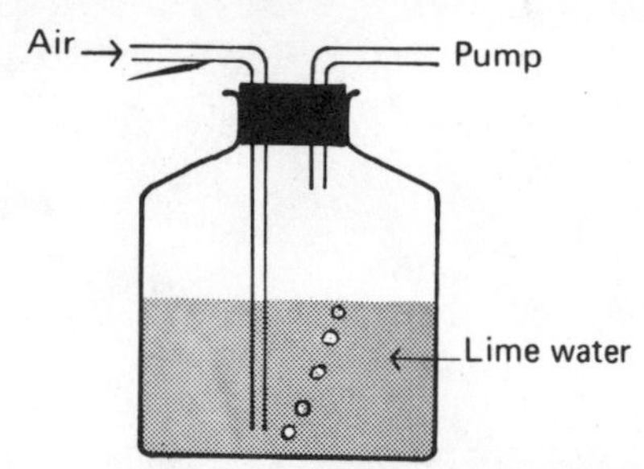

Fig. 21.4. Carbon dioxide in air.

How long did it take for the limewater to go faintly cloudy?

There is not much carbon dioxide in the air and it is very hard to measure it. But scientists have found out that in 3000 cm³ of air there is only 1 cm³ of carbon dioxide. Look at the 5 litre bottle again. This has less than 2 cm³ of carbon dioxide in it.

Nitrogen in the air

The air contains $\frac{1}{5}$ of oxygen, $\frac{1}{100}$ of water vapour, $\frac{1}{3000}$ of carbon dioxide. If you add these up you

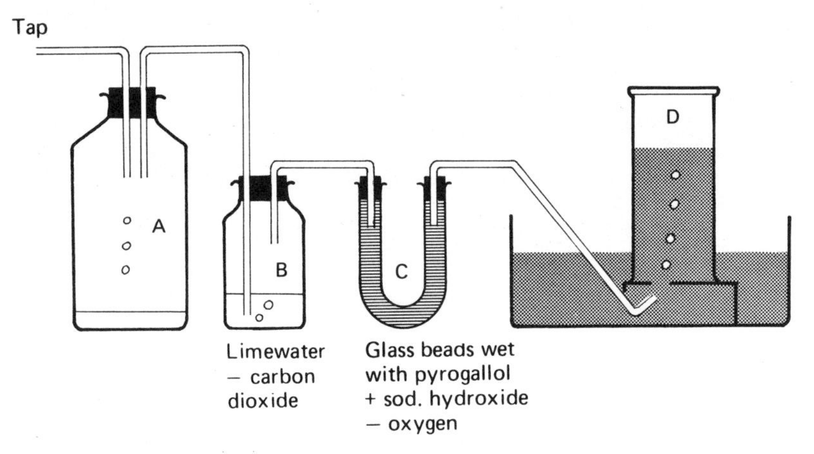

Fig. 21.5. Getting nitrogen from air.

will see that this still leaves about $\frac{4}{5}$ of the air which must be something else. To get this $\frac{4}{5}$ we must take away from air its oxygen, water and carbon dioxide. What can we use for taking away the oxygen? The carbon dioxide? (We won't bother to take out water because it will get wet again as we collect it.) The apparatus is shown in Fig. 21.5.

At the start, bottle A has air in it. By pouring water into it we can make air go through the apparatus. In B carbon dioxide is taken away, and in C the oxygen is taken away. We collect what is left in D. What do we call this way of collecting a gas? Bubble air through very slowly, about one bubble every two seconds. When you have filled D, put a slide over it and take it out. Put a lighted splint into it.

What happened to the splint? Why?

Fig. 21.6. This job must be done in clean air so that no dust gets onto the parts. The cabinet is in fact filled with nitrogen gas rather than air because any oxygen would damage the very clean surfaces that are being put together. Is the lady wearing gloves? Can nitrogen escape up her sleeves?

This $\frac{4}{5}$ may not seem very exciting and we cannot do many experiments with it. But scientists have found that it is a mixture. There are several elements in it. (What is an element?)

Almost all of it is *nitrogen*, but it also has small amounts of *argon*, *neon*, *krypton* and *helium*. You may have heard some of these names before.

The air does not have fixed amounts of gases in it, but it is useful to have a rough idea of what is there.

Name	In every million cm^3 of air there are
Nitrogen	780 000 cm^3
Oxygen	200 000 cm^3
Water vapour (varies a lot)	10 000 cm^3
Carbon dioxide	300 cm^3
Argon	10 000 cm^3
Neon	10 cm^3
Helium	3 cm^3
Krypton	1 cm^3

The noble gases

Argon, neon, krypton and helium were first discovered about 1900 (most elements have been known a lot longer). They are called the 'noble gases', and you may wonder why this is. What is 'noble' about them? For many years before their discovery, gold and silver were called 'noble metals'. Metals like iron would rust away in time, gold and silver would not, so chemists thought how 'fine and noble' they were. Coins were made of them because they would not rust away; they stayed the same for a long time. Argon, neon, helium and krypton are like gold and silver in a way. They too are inert, so we call them the noble gases.

Fig. 21.7. Are airships the answer to cheap long-distance air travel? They are filled with a very light gas, but hydrogen is so very dangerous – why? Why is helium preferable to hydrogen?

Uses of the noble gases

You may think that gases which will not react with things cannot be very useful. But this is not true; the fact that they are inert *makes* them useful for many things.

Weather forecasters get information by sending balloons into the upper atmosphere. Many years ago hydrogen balloons were used because the gas is so light – it has a small density. (What does this mean?) But do you remember the test for hydrogen? Think what might happen if a light came near to the balloon. Nowadays helium has taken the place of hydrogen. It is not so light but it is safer *because* it is inert.

Argon is used for filling electric light bulbs. What kind of energy goes into a bulb? What is it changed to? What might happen if the filament in the bulb had air round it? Which gas might combine with the filament? The metal would become a powdery oxide and break. But if the bulb is full of argon this cannot happen.

You may have seen the illuminated signs on big shops. They flash on and off and they have lovely colours. The reddish orange one is filled with neon gas.

More about molecules

You now know the names of a few more elements, but what are their molecules like? The nitrogen molecule has a pair of atoms linked together.

Fig. 21.8. This is a night scene over New York. The light radiates from bulbs which are full of rare inert atmospheric gases.

Fig. 21.9. This is a welding machine which melts metal by making electric sparks. As it makes the metal very hot the air must be prevented from rusting the metal; this is done by forcing argon out of the nozzle so that 'the work' cannot come in contact with oxygen. Why is the fitter wearing a shield and gloves? What tubes must run through the handle of the welding machine?

We can draw the gas with the molecules speeding through space, but the atoms stay linked together.

Can you think of another gas whose molecules are like this?

The noble gases are very different. Their atoms are not linked together and they move on their own. We can say that the molecule has one atom in it.

A substance whose molecules have two atoms in them is called *diatomic*; one whose molecules only have one is called *monatomic*.

As you would expect from the name, carbon dioxide is made up of molecules which have carbon and oxygen atoms in them. Here is a molecule of the gas.

(O)——(C)——(O)

Like water it is *tri*atomic.

Summary

About $\frac{4}{5}$ of the air is an element called nitrogen and this is inert.

About $\frac{1}{100}$ of the air is water vapour and this can be taken out by using deliquescent chemicals. These take in water and get wet.

Air has a little carbon dioxide in it, and we can take this out by using limewater.

Other inert gases are argon, neon, krypton and helium, and we call these the 'noble gases'. They are all elements. Together they make up $\frac{1}{100}$ of the air.

Diatomic molecules have two atoms in them, triatomic three, and monatomic one atom.

Questions

(1) Give the names of five gases which are in the air, and give the amounts of three of them.
(2) What is meant by 'deliquescent chemical'? Write the names of three substances which are like this.
(3) What is an inert chemical? Give the names of one metal and three gases which are like this.
(4) Give three uses for the noble gases.
(5) What is a monatomic substance? Give two examples. What is a diatomic substance? Give two examples. What is a triatomic substance? Give two examples.

Something to do at home

(1) If you made a microbalance last year – why not use it to find things which take in water from the air. Protect the straw with a small cup made of silver paper and put the material in this. Can you find any substances which lose weight?

(2) Nitrogen is inert – but it does have one or two reactions. Find out in the library how it combines with oxygen, hydrogen and magnesium. Write about four lines on each reaction.

Chapter 22 More about Oxygen

Fig. 22.1. Apollo 10 blasts off! The chemical energy in liquid oxygen and paraffin are changed into motion energy. The space vehicle and the men inside it gain enough speed to reach the moon. They use hundreds of tonnes of fuel in the space of a few minutes and to get enough in the rocket the gases have to be stored as liquids. Find out the temperature of liquid oxygen. As the rocket speeds upwards what becomes of all this chemical energy?

We have already found out that oxygen is a very important gas. We use it when we breathe or when we burn things. Many substances are anxious to combine with it so we can expect to find it in a lot of different compounds. What do we call a compound which has oxygen and one other element in it? What other word endings tell us that a compound has oxygen in it?

If we heat a compound containing oxygen we may be able to break it up and collect the oxygen. We will try to do this now. (What other kinds of energy are able to decompose chemicals?)

Make a list of chemicals which contain oxygen in your note-book, and then decide with your teacher which ones you will work with. Write their names on small pieces of paper – a quarter sheet for each one – so that you are ready when the chemicals come round.

Put some of the first chemical into an ignition tube and heat it. Remember the rules for heating things. Every few seconds test for oxygen – it will help you if the splints are made into even narrower strips.

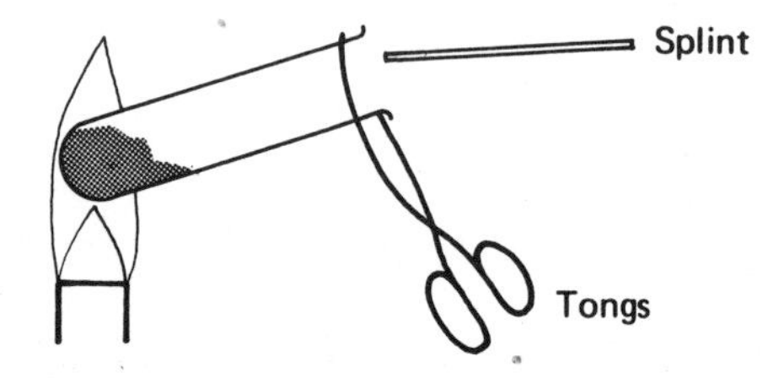

Fig. 22.2. Getting oxygen by heat energy.

Make a list in your note-book and fill it in:

Name of chemical	Was oxygen given off	Anything else of interest

Now answer the questions:

Did they all relight the splint? Which did it the most easily? Did any give off steam? Do you think that steam may stop the splint from relighting if oxygen is there?

Which would be the easiest to use for making some oxygen?

It is not always convenient to heat things up, and it would be easier if we could get the oxygen without using heat. We know that some chemical changes start without heat (give one which does), and that some start if a catalyst is there. We will now try to get oxygen in this kind of way.

Oxygen from hydrogen peroxide

Hydrogen peroxide is a very useful chemical. Some people mix it with water (dilute it) to make a gargle. It is harmful when it is strong, but when it is dilute it will clean out your throat and mouth very thoroughly. When it is very strong it can be used as a rocket fuel.

The 'peroxide' in the laboratory is just like the liquid you buy at the chemists, but it is probably stronger. You will notice that it is kept in brown bottles. Why is this? A lot of chemicals are able to make hydrogen peroxide give its oxygen quickly – even dust will do it, but we will use manganese dioxide as the catalyst.

When a chemical breaks up easily we say that it is *unstable*. Things like dynamite and T.N.T. are very unstable indeed, and they decompose violently.

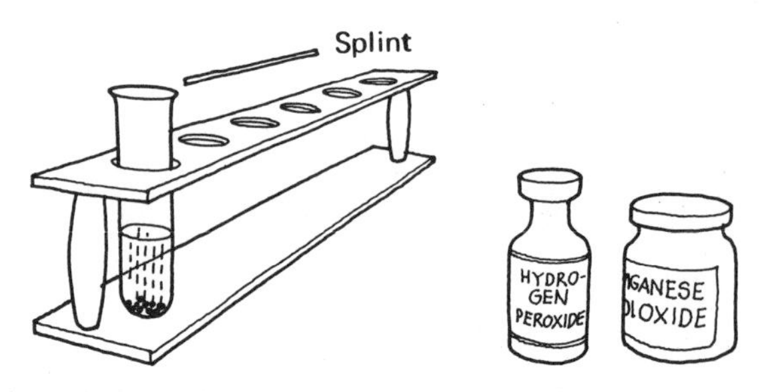

Fig. 22.3. Making 'peroxide' give its oxygen.

Take a clean test tube and pour in 1 cm of hydrogen peroxide. Now add a spatula of manganese dioxide and test for oxygen. But now we want to collect the gas so we will do the same experiment but change the apparatus (see Fig. 22.4).

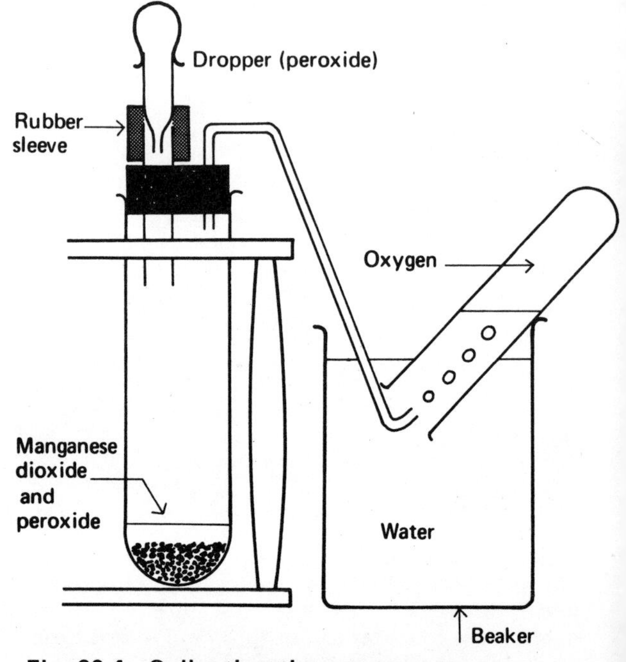

Fig. 22.4. Collecting the oxygen gas.

Start with the bottom of the boiling tube covered with manganese dioxide and add the 'peroxide' using

a dropper. As one tube fills change it for another. Collect 4 tubes of gas.

Now that we have some tubes of the gas we can find out interesting things about it. Here is a list of things we want to know; answer the questions in your note-book under a heading 'the properties of oxygen'.

(1) Can you smell it?
(2) Can you see it?
(3) Does it let things burn in it?
(4) Does it burn?

(1) When something has no smell we say it is *odourless*.
(2) When a gas is invisible we say it is *colourless*. Use these words in your note-book.
(3) and (4) Give the names of two things which burn well in oxygen, apart from wood. Oxygen allows things to burn in it, and we say that it *supports combustion*. But does it burn itself? If you put a light to your Bunsen burner the gas burns. But if you put a light to oxygen and then take the light away – does the oxygen burn? We say that oxygen is not *inflammable*.

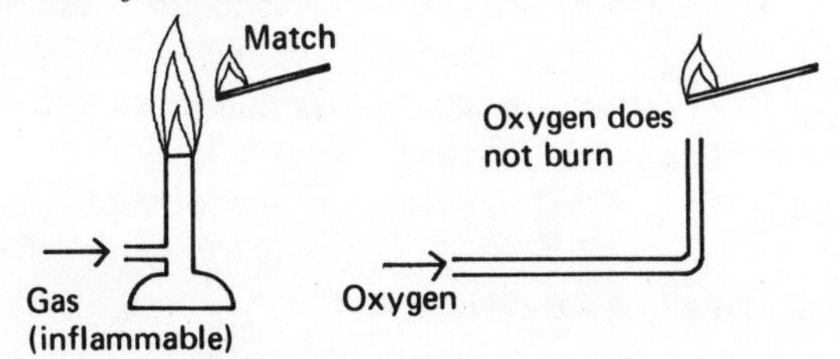

Fig. 22.5. Oxygen supports combustion but does not burn.

Burning elements in oxygen: making oxides

(1) *Does iron burn?*
If you want to make a wood fire, you start by lighting small pieces. Why do you do this? We will try to make iron burn, but it would be silly to start with a bar of iron so we will start with fine strands of iron called iron wool.

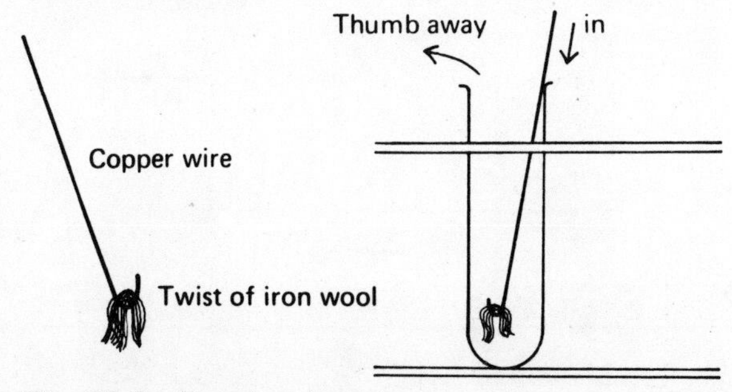

Fig. 22.6. Burning iron.

Take a tube of oxygen out of the water and put it in a test tube rack keeping your thumb over the end – why? Get your partner to twist some iron wool round a hook made of copper wire (see Fig. 22.6). Start the wool glowing in a Bunsen and quickly move it into the tube of oxygen. Keep your hand away from the tube when the iron is put in.

Start off a report in your note-book:

Element burned	What I noticed	Chemical made

(2) *Does carbon burn?*
What will be made if carbon burns in oxygen? How do you test for this compound? Ordinary charcoal is carbon and we will use this.

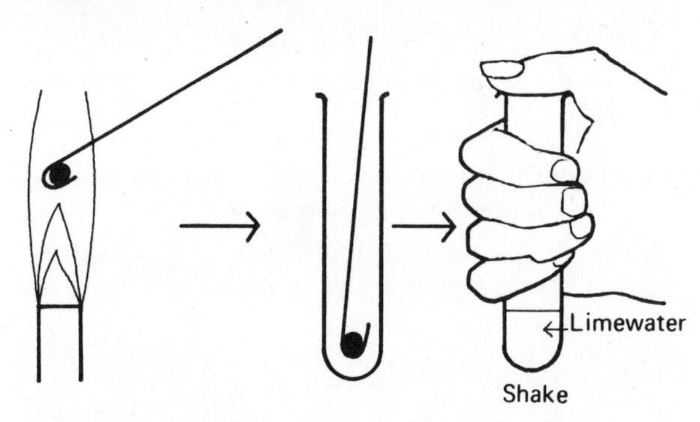

Fig. 22.7. Burning carbon.

Fold the copper wire round a piece of charcoal and heat it till it glows. Quickly move it into a tube of oxygen. As soon as the reaction stops, take the wire out and pour in a little limewater. Gently shake the tube. Fill in the table in your note-book.

(3) *Sulphur*
First get an asbestos square to protect the bench. You will be given a piece of wire with some sulphur on one end of it. (How was it made? What happens to sulphur when it is heated?) Heat the end so that the sulphur starts to burn (you will have to look carefully but don't put your face too close) and put it in a tube of oxygen. Now fill in the table. Notice how unpleasant the fumes are, but don't sniff them too hard.

Chemical names
What is the name of the chemical made when iron burns in oxygen? When magnesium burns in oxygen? When carbon burns it gives carbon *di*oxide, and sulphur gives a gas called sulphur *di*oxide. Do these names tell you anything about

their molecules? A molecule with one oxygen atom in it is called a *mon*oxide; one with two is called a *di*oxide; and one with three is called a *tri*oxide. Mono, di and tri mean one, two and three. How many wheels does a tricycle have? What is a monorail? You can see pictures of some molecules in Fig. 22.8. If the symbol for the manganese atom is Mn, draw a molecule of manganese dioxide.

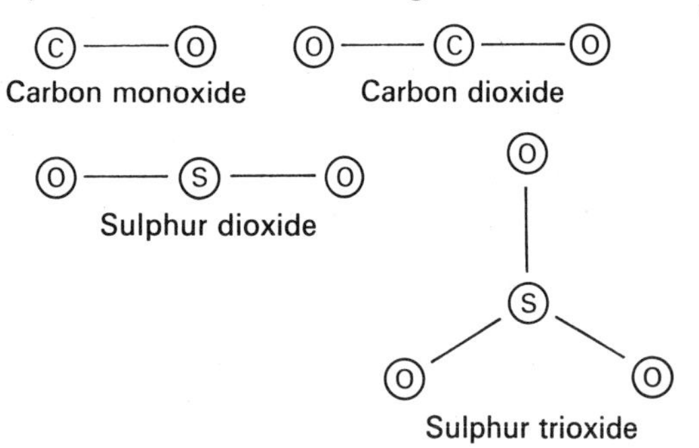

Fig. 22.8. Some molecules with oxygen atoms in them.

You may think it strange that carbon and oxygen make two different compounds. But is it really so surprising? When your mother bakes short bread biscuits she mixes one part of flour with one part of fat. But when she makes pastry she uses two of flour to one of fat. Making chemicals is like this. One atom of carbon and one of oxygen give a molecule of carbon monoxide; one of carbon and two of oxygen give a molecule of carbon dioxide. Of course in cooking you can put any amounts of fat and flour together (you probably do at camp!). But when you make molecules you cannot do this because atoms cannot split up.

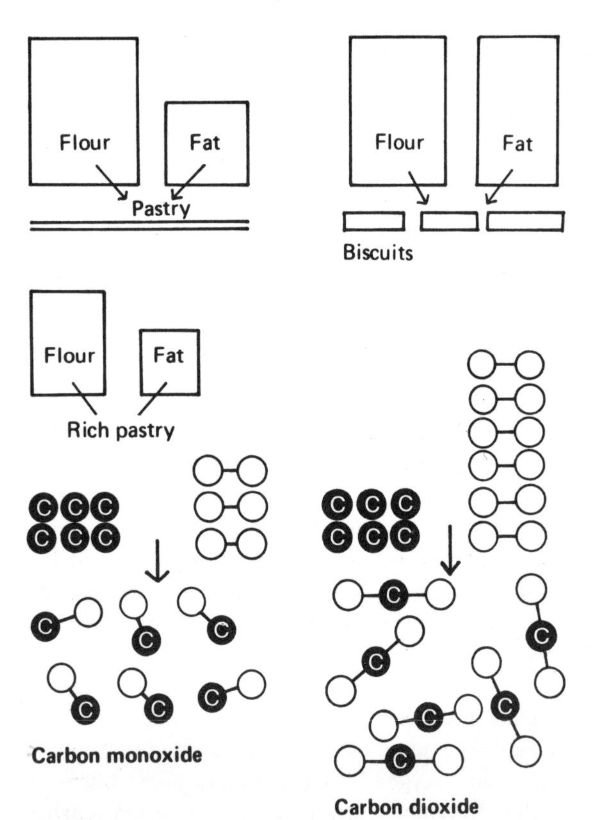

Fig. 22.9. Different compounds from the same elements.

An important kind of chemical change

What fraction of the air is oxygen? You will see that there is a lot of oxygen gas about, and many substances get a chance to combine with it. It is so common for oxygen to combine with other things that we have a special name for the reaction. When a substance combines with oxygen we say it has been *oxidised.*

Sometimes we say that an *oxidation reaction* has happened.

Here are some oxidation reactions and in each of them the first substance has been oxidised. The arrow is used as a shorthand way of saying 'gives'.

Carbon + oxygen ⟶ carbon dioxide,
copper + oxygen ⟶ copper oxide,
sodium + oxygen ⟶ sodium oxide,
carbon monoxide + oxygen ⟶ carbon dioxide,
wax + oxygen ⟶ carbon, water, and carbon dioxide.

Something to do at home

See if light will decompose hydrogen peroxide to give some oxygen. Do the experiment on a window sill – it may take weeks (Fig. 22.10).

Fig. 22.10.

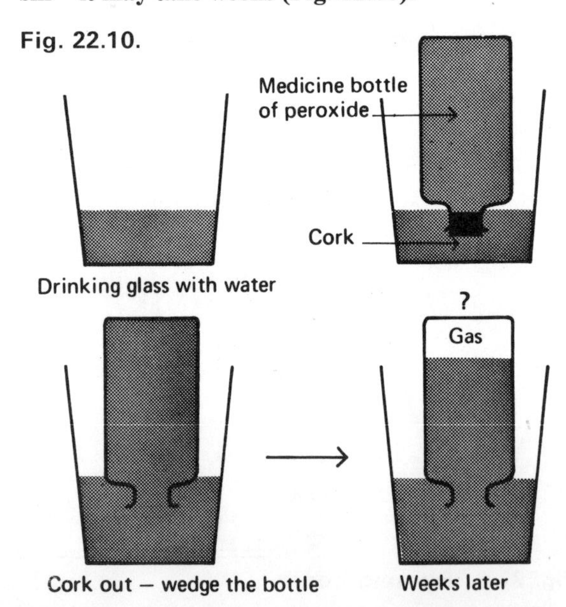

Questions

(1) Write the names of four chemicals which give oxygen when they are heated.

(2) Name four elements which burn in oxygen. Name three things which are not elements and which burn in oxygen.

(3) Oxygen supports combustion – what does this mean?

(4) Draw a diagram of the apparatus which you used to make oxygen.

(5) Write down the names of the substances formed when the following combine with oxygen: copper, zinc, magnesium, carbon, hydrogen, sulphur dioxide, sodium.

(6) Draw molecules of the following compounds: nitrogen dioxide, lead dioxide (Pb for lead atom), chromium trioxide (Cr for chromium atom), lead monoxide, aluminium trichloride (Al), carbon disulphide.

(7) Name four gases which do not burn and which do not support combustion.

(8) Write an essay about fire extinguishers – drawing diagrams of them, telling what is inside them, and explaining why they put fires out.

Chapter 23 Light Energy

One of the kinds of energy that we met in Chapter 13 was light energy. What kind of changes can light energy bring about? Daylight comes to us from the sun and so it must be able to travel through empty space. In fact light is a special kind of radiation energy and we are provided with special parts of our bodies to detect it – eyes. Our eyes contain substances that change when the light energy falls on them, and this makes us able to see things.

Fig. 23.1.

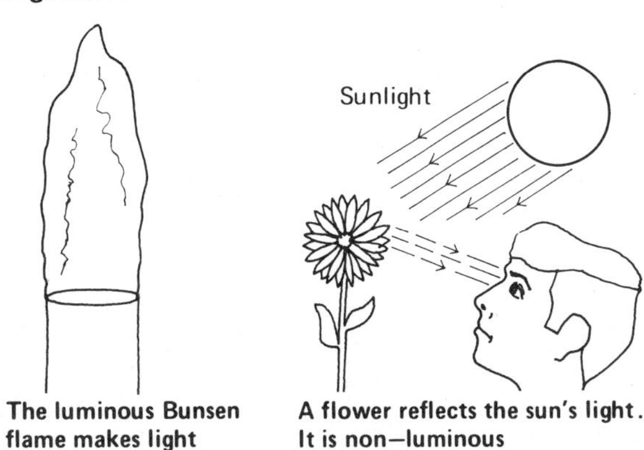

The luminous Bunsen flame makes light

A flower reflects the sun's light. It is non–luminous

When we look at something, light energy travels from the thing to our eyes. Some things – a Bunsen flame, for instance, make their own light and they are called *luminous*. Other things, such as flowers or coloured dresses, don't make light themselves but they are able to reflect the light from the sun or from a lamp into our eyes. We call them *non-luminous* objects. Some substances such as water and glass allow light to travel through them and we say they are *transparent*. Others, like wood, metal or pottery, do not, and we say they are *opaque*. You can't see things through an opaque substance. We are going to try to find out more about how light travels from one place to another.

At some time or other you must all have seen sunlight shining through a gap in the curtains and crossing the room as a band of light until it falls on a wall or floor in a bright patch of light. If the room is a little dusty or smoky the path of the light is clearer and we see that it travels in straight lines across the room. We call a band of light of this kind a *beam*. If it is a very narrow beam we call it a *ray*. The arrow we put on a ray shows the way the light is going, not the way you are looking when you see it.

Fig. 23.2. Light travels in straight lines. Searchlights like these are used in defence against raiding aircraft.

Something to do at home

(1) Light an electric torch and let the light fall on a wall. Look carefully at the patch of light so that you can recognise the patch made when your torch beam hits a surface. In a darkened room shine the beam onto a polished table or other shiny surface. You will see a patch of light on the table. You will also be able to find somewhere a fainter patch of light that you can see comes from your torch. As you move your torch about so the patches move about. You may see more than one fainter patch of light, if so try to work out how the light from your torch gets to them. A glass mirror gives specially interesting effects.

Fig. 23.3. An image made by the reflection of light at the surface of a pool of water. Is the image a real image or a virtual one? Measure the width of the dome and of its image. Is the image magnified? In what country does this building stand, and what is it called?

When light bounces off a surface in this way we say that it has been *reflected*.

What surfaces reflect light best, polished surfaces or dull ones? Does the colour of the surface make any difference to the reflected patch? Does the patch look the same if it has been reflected by a curved surface like the side of a glass jam jar?

(2) Put a polished surface – a mirror or a piece of glass is best but a polished table will do – in a place where you can use it to reflect the torch beam onto a wall as in Fig. 23.4. Stand a pencil on the table – or use a knitting needle stuck upright in a cork or a piece of wood. Shine your torch so that the pencil is roughly in the middle of the patch of light it makes on the table, as in Fig. 23.4 (*a*), and notice whereabouts on the wall the reflected light comes. Now shine the torch down more steeply, as in (*b*), and

Fig. 23.4.

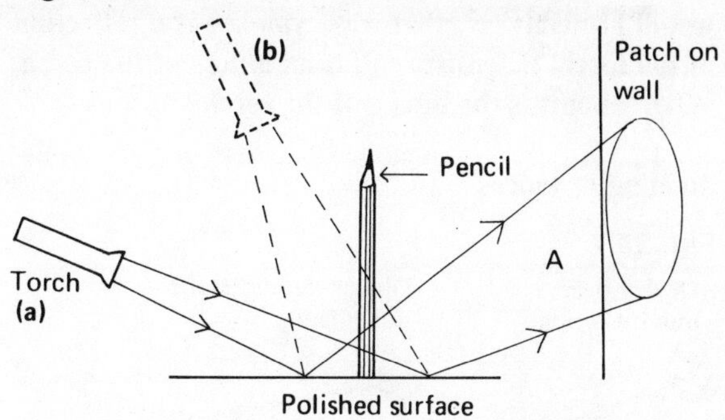

watch what happens to the patch on the wall. Does the reflected beam come off the surface more steeply or less steeply?

The pencil, or upright marker, or any line at right angles to the surface we call the *normal*. (This word has quite a different meaning in daily life.) The angle between the normal and the torch beam we call the *angle of incidence*, and the angle between the normal and the reflected beam we call the *angle of reflection*. What happens to the angle of reflection if you make the angle of incidence bigger?

Fig. 23.5.

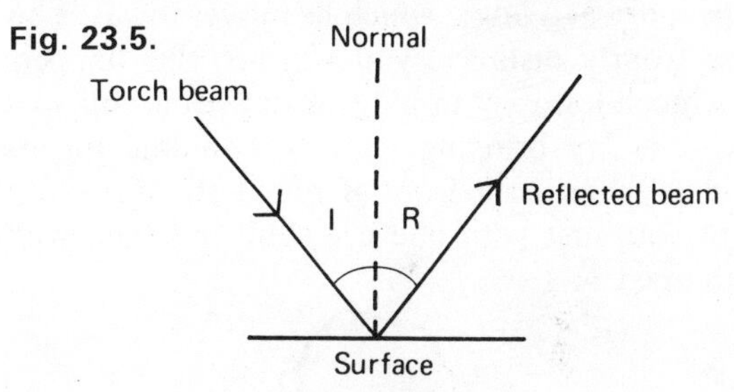

(3) Use a really shiny surface – a mirror or a bright metal – for this experiment. Prop your torch up so that it shines on the surface as in Fig. 23.6. If you have an adjustable reading lamp you can use it instead of a torch. Move your head so that you are looking at the mirror from A, near the patch on the wall. Where does the light seem to be coming from? Hold a book or a piece of paper so that you can't see the torch itself.

Fig. 23.6.

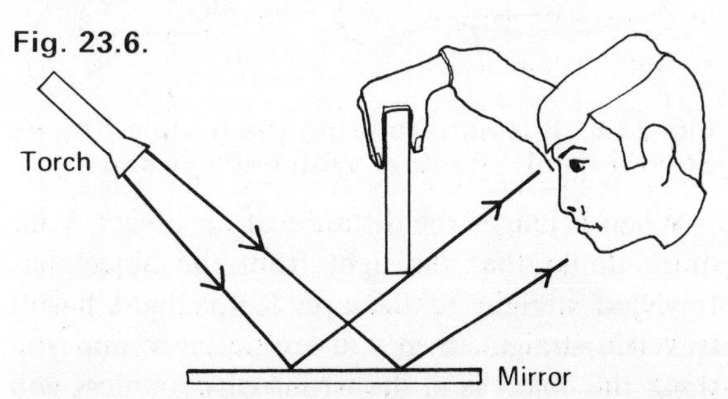

If you didn't know where the torch was, where would you think it was – in front of the mirror, on

the mirror, or behind the mirror? In daily life you would probably say that what you see is a reflection of the torch. Scientists call it an *image* of the torch. Whereabouts is the image of the torch?

Judging distances

Fig. 23.7.

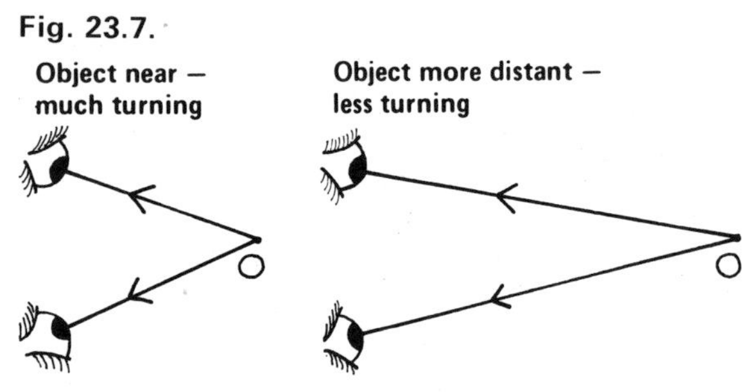

When you look directly at something the light from it travels straight to your eyes and so you see it. Because you have two eyes the light reaches them from two slightly different directions and your brain judges how far from you the thing is.

Your eyes turn so that both of them point to the object you look at. If you watch the eyes of a friend as he looks at a finger which he moves towards his nose from a distance, you will see this happen. It is much less easy to judge distances if you shut one eye. Try bringing your two middle fingers suddenly together in front of you at about two feet from you, first with one eye shut and then with both open.

Fig. 23.8. It is hard to bring the middle fingers of your hands together with one eye shut.

When it judges the distance of an object, your brain thinks that the light from the object has travelled straight to the eye. If the light hasn't travelled straight, then you are deceived and you think the object is in the wrong place, unless you happen to *know* where it is. If you have a young puppy or kitten at home you can have a lot of fun

Fig. 23.9.

by just standing a mirror on the floor against a chair and watching what happens when the kitten catches sight of its image. You can see how it is deceived. It soon learns, though! The experiment doesn't usually work more than once with the same kitten!

Fig. 23.10.

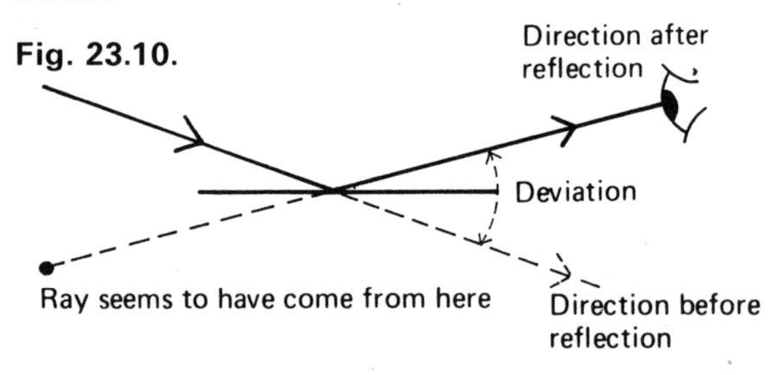

When a ray of light is turned aside from its straight line path by a mirror or any other thing we say it has been *deviated*. It is when rays of light are deviated that images are formed, and the eye may be deceived.

Something to do at home

Deceiving your eyes

Fig. 23.11.

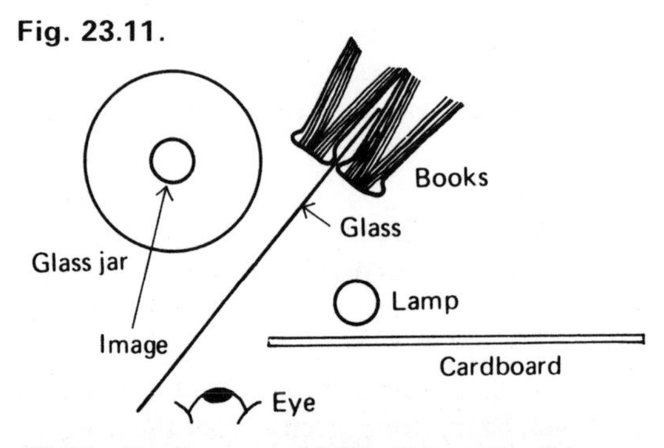

(1) Use books or wood blocks or plasticine to prop up a sheet of glass as in Fig. 23.11. Put a flash-lamp bulb in front of the glass as shown and look at the image of the bulb from roughly where the eye is

drawn. Put a piece of cardboard or plywood so that you can't see the lamp itself. Stand a jam jar half full of water behind the glass and move it about until it covers the image. It looks now as if the bulb is under the water. It is still more mysterious if you use a candle instead of a bulb, but then you must be careful not to drop grease about. If you are good at woodwork you can make a wooden box to hold the jar, glass and candle so that people can't see how the trick is done.

This arrangement is called Pepper's Ghost, after a man who first used it on the stage to deceive the audience.

(2) If you have two pieces of mirror and some plasticine you can arrange to see round a corner or over a wall. Arrange them as shown in Fig. 23.12. The torch beam will be reflected twice and if W is a wall you will be able to look over it without exposing your head. The angle marked must be 45° and the two mirrors must be parallel. You can support them in two slots in a wooden board as shown in (*b*).

Fig. 23.12.

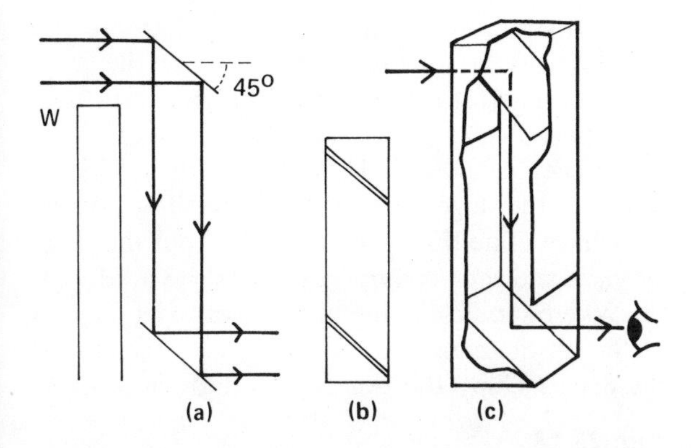

When you have the arrangement in working order you may be able to make a box to hold the mirrors, as in (*c*). You then have a home-made *periscope*. Periscopes are used in tanks and submarines so that people can see without being seen.

Experiment: *More about the angles of incidence and reflection*

Fig. 23.15.

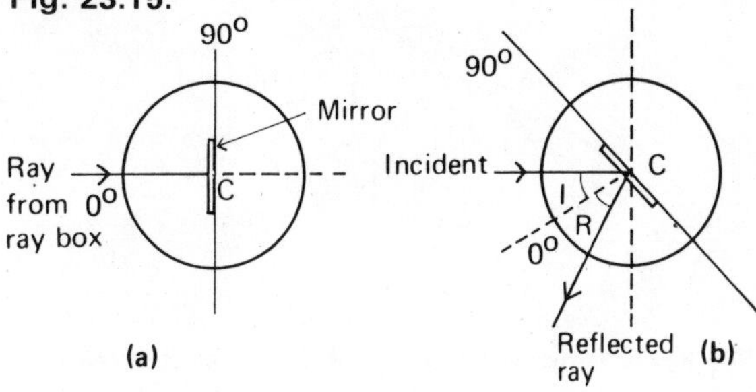

You will be given a ray-box – a means of getting a fine beam or ray of light. You will also need a *protractor*, which is a circular paper scale with angles marked on it. Put the protractor on the bench as in Fig. 23.15 and on it prop a piece of mirror so that the back of the mirror is along the 90° line. Make sure that the ray hits the back of the mirror right in the centre C of the scale, as shown. Turn the mirror a few degrees as in (*b*) keeping the ray at C. The ray before it reaches the mirror is called the *incident* ray. What are the names of the angles I and R marked in (*b*)? Write down in your book the sizes of these two angles. Move the mirror, bit by bit, and repeat the measurement each time. What do you discover about the angles I and R?

Fig. 23.13. The officer is looking into the bottom end of a periscope. Fig. 23.14 shows what he sees. The naval periscope has lenses built into it so it gives a magnified image.

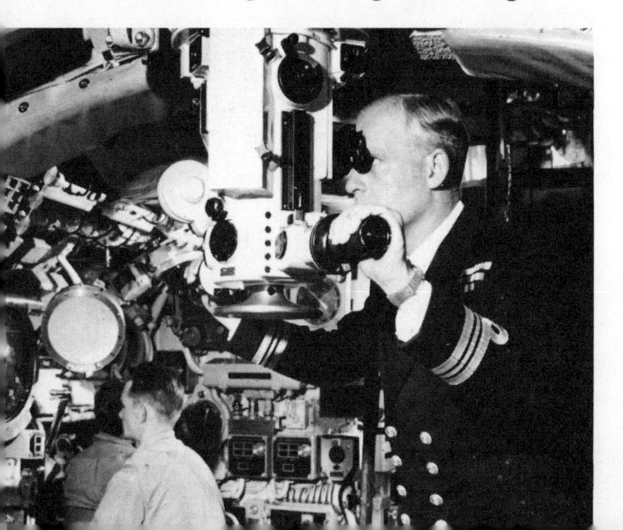

Fig. 23.14. How things look through a periscope. This is a view of Saint Peter and Saint Paul Rocks, in mid-ocean between South America and Africa, as seen through the periscope of a submarine.

Which part of the mirror reflects most light? Why is the *back* of the mirror placed along the 90° line? Look again at Fig. 23.12. Why must the angle marked be 45°?

Something to do at home

Fig. 23.16.

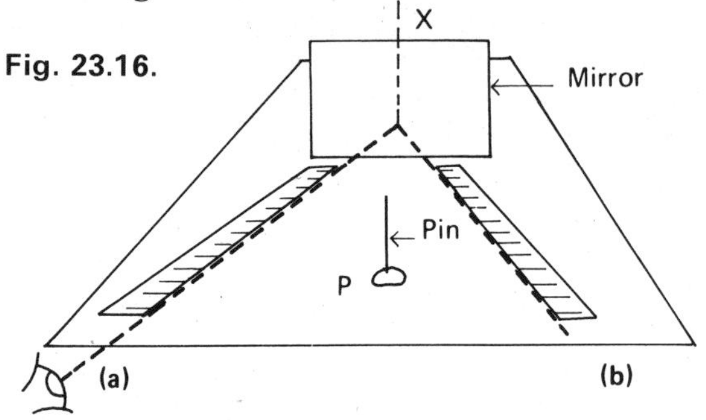

We know from our Pepper's Ghost experiment that the image of a thing is somewhere behind the mirror that makes it. You can find out more about this using the arrangement in Fig. 23.16. Rule a line across the middle of a piece of paper and prop a small mirror up along the line. A lady's vanity mirror will do quite well. Will you put the back or the front on the line? Mark a point in front of the mirror and just over P put a pin stuck in a piece of cork or plasticine or an indiarubber. Or use a mounted flash-lamp bulb if you have one. Looking from one side, (*a*), put a ruler on the paper so that its edge points at the image. Hold the ruler steady and pencil along the edge. Do the same from the other side, (*b*). Take care not to move the pin or mirror while you do this. Remove the mirror and pin and continue the two lines until they meet, at X say.

From which direction does the light seem to come when you look from (*a*)? from (*b*)? Where would you suppose the light came from if you really didn't know? Mark on your piece of paper how light *really* got to your eye. How far is P from the nearest point on the mirror line? How far is X? Is one distance much bigger than the other? Are they roughly the same? Exactly the same? Would this be true wherever you put P or from whatever place you looked? Try it a few times!

A home-made raybox

You can easily make up your own raybox using the items shown in Fig. 23.17, which is very easy to understand. The lamp can be a flash-lamp bulb – try propping an electric torch so that it points at the jam jar. Also try removing the glass from the front of the torch – sometimes it works better that way. Or you can use a separate bulb and a dry battery.

Fig. 23.17.

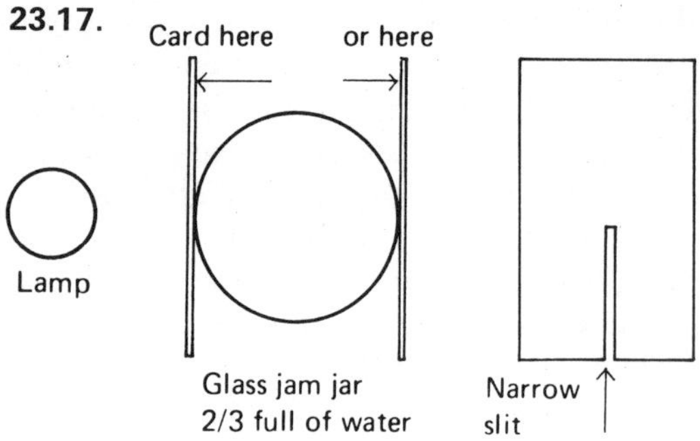

Card here
or here
Lamp
Glass jam jar
2/3 full of water
Narrow
slit

You should adjust the height of the torch and its distance from the jar until you get a good parallel beam of light coming out of the jar. Then place your slit in either of the places shown – whichever works best with your gear – and produce your ray.

You can of course cut two, three or many slits in the card and so study what happens to the various rays when they are reflected. You can also try using the teeth of a comb to make your rays and perhaps other good ideas will occur to you.

Questions

(1) Can you hold a single mirror so that you can see the back of your head? Can you see it if you have two mirrors? If so give sketches to show how it is done.

(2) Dr Pepper used a sheet of glass to produce a ghost on the stage. Draw a sketch to show how he may have done this; show where he put the sheet of glass and the person whose ghost he produced.

(3) Where should a man put a mirror to let him see traffic on the main road as he reaches the end of his drive? Copy the sketch and add the mirror.

Fig. 23.18.

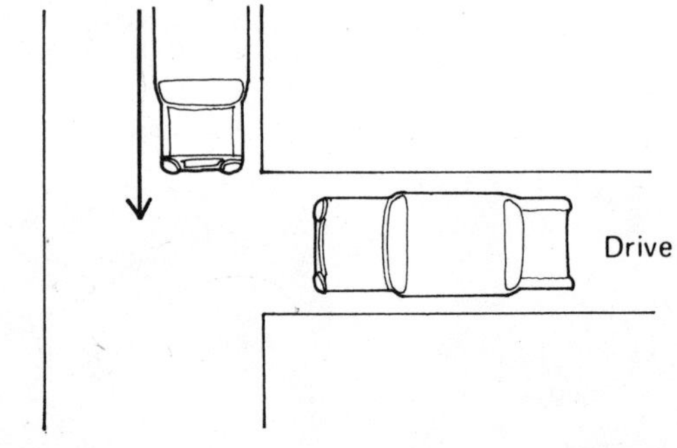

At what angle must the mirror be placed? Will one mirror be enough?

(4) Draw a sketch to show how a dentist can use his mirror to see the back of one of your teeth.

(5) What is a 'kaleidoscope'? If you have three strips of mirror or pieces of polished metal you can make one for yourself.

Images without mirrors

We now know that our eyes can be deceived by hidden reflectors that deviate the light rays. There are other ways of making light rays deviate without using mirrors. Pour water into a wash basin and then stand a ruler or stick in the water and look at it carefully. Your eye is certainly being deceived again and the water surface seems to have something to do with it.

Experiment (You can do this at home if you have a home raybox and a flat medicine bottle with a cork)

Fig. 23.19.

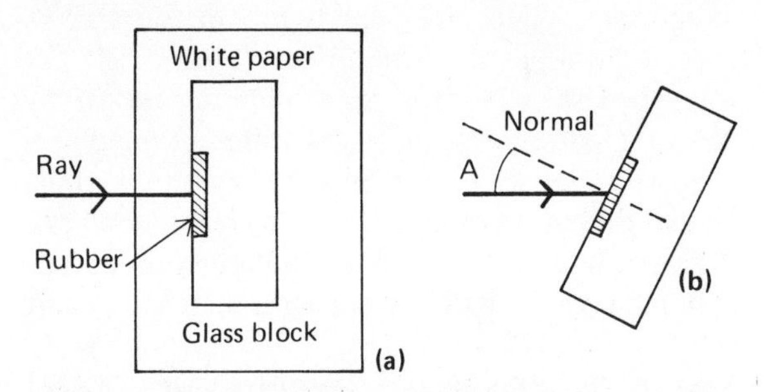

You need a block of glass – or you can use a medicine bottle filled with water (it is best to wire the cork firmly in place). Arrange the apparatus as in Fig. 23.19 so that a ray from your ray-box falls on one side of the block as in (*a*). Look down at it. What happens to the ray? Turn the block so that it takes up the position as in (*b*). We stand the rubber (or plasticine) on the block because we are interested in the light that goes through the block and not over the top.

Does the light go straight into the block, or does it change direction? in (*a*)? in (*b*)? If not, does it bend towards the normal or away from it as it goes in? Does all the light at the bottom go into the glass? If not, what happens to the rest of it? Is any light reflected at the back surface of the glass? Does the light come straight out of the back of the block in (*a*)? in (*b*)? If not, does it bend towards the normal or away from it as it comes out? What do you notice about the incident ray and the ray that comes out from the far side of the block?

Make a sketch in your book showing the path of the light through the block in (*a*) and (*b*).

You will have noticed that as the light goes from air into the block it is bent one way, and as it comes out it is bent the other way, the same amount. The surface of a mirror deviates a ray by reflection. The deviation that happens when a ray crosses the frontier between two different substances is caused by what scientists call *refraction*. When light passes from water into air it is refracted. This is why the stick in the basin appeared bent. Fig. 23.22 shows what happened.

Fig. 23.20. Refraction. The light ray bends as it goes into glass and again as it comes out. Notice the direction of the ray as it comes out. Has it been deviated? There are other fainter rays in the picture. How are they produced?

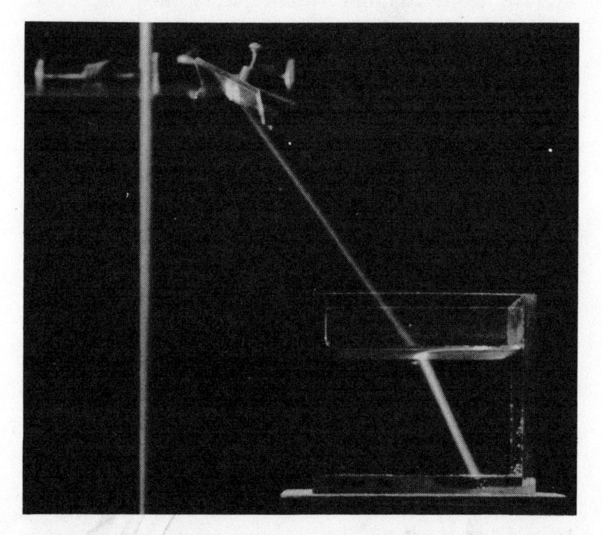

Fig. 23.21. Refraction. A water surface will also bend light rays. Look again at Fig. 23.20.

Fig. 23.22.

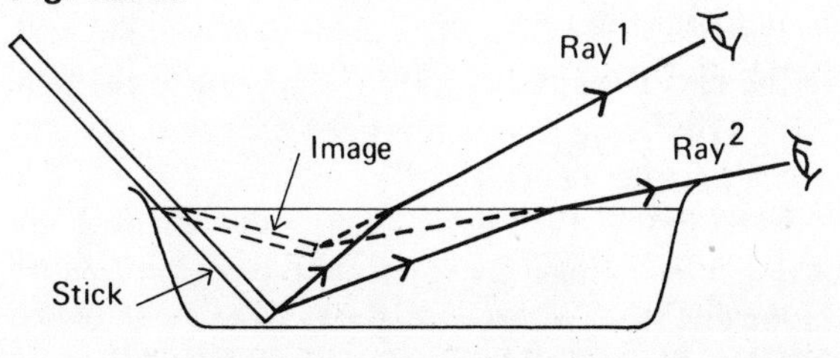

Rays 1 and 2 are refracted as they come out of the water and so the eyes of the observer are deceived and he thinks that they have come from somewhere higher up.

Something to do at home

The mysterious penny

Fig. 23.23.

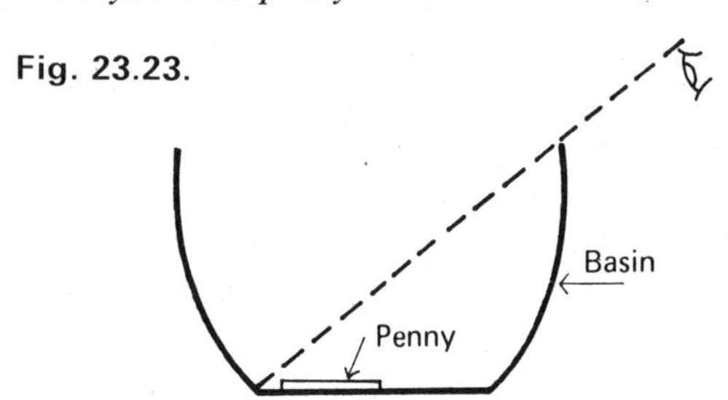

Take a pudding basin or something of the kind and put a penny in it as shown. Seat yourself so that you just can't see the penny over the side of the bowl. Get someone to pour water gently into the bowl, so as not to disturb the penny. After a while you will be able to see the penny. Copy Fig. 23.23, draw in the water surface, and mark in how rays can now get into your eye from a point on the penny. (You need not copy the dotted line. Why is it put in the figure?)

Experiment

Fig. 23.24.

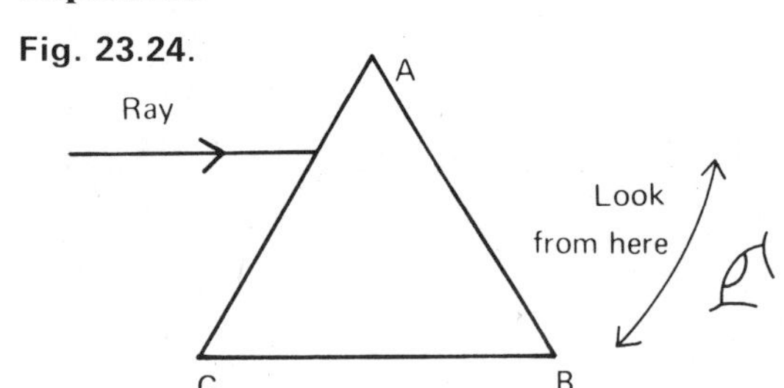

You will be given a triangular shaped piece of glass or plastic called a *prism*. Stand it on a sheet of paper and arrange your ray-box so that a ray falls on one side of it roughly as in Fig. 23.24. If you look down on the prism, you will be able to see the light inside the prism. Is the ray deviated as it goes into the prism? Does the ray inside the prism reach AB before BC? Does all the light energy go into the prism? Does the ray go through AB into the air?

Turn the prism to and fro and watch what happens to the ray. Does the ray always get out of the face AB? What happens to the light energy if the ray doesn't get out of AB?

Arrange the prism so that light is coming out from AB and look along the ray from the position shown in the diagram. Where does the ray-box seem to be? Mark it on a sketch. Has the prism made an image of the lamp? Is it a good clear image? The light from the ray-box is ordinary light, but the light coming out of AB is coloured. Write down what colours you can see (look carefully!).

Take a piece of red glass or red cellophane and hold it between the ray-box and the prism. Can you still see all the colours? If not, which colours are missing? Which are left? What would you expect to see if you used a piece of blue glass? Try it if you can.

Three hundred years ago Sir Isaac Newton did this simple experiment and many others to find out why the light turned coloured when it passed through the prism. We now believe that all the colours are present in the white light to begin with – that white light is a mixture of many colours. Each colour is deviated a little differently by the prism, so that they come out in slightly different directions. When red glass is put in the way of white light it stops all the colours except red. The stained-glass windows in a church appear in different colours because the different pieces of glass subtract different colours from the daylight and so different colours are left to reach your eye.

Sometimes reflection makes light coloured. You may have seen a patch of red light on a wall where

Fig. 23.25. Sir Isaac Newton (1642–1727). One of the very greatest scientists. He worked first at Cambridge and later became Master of the Royal Mint. His most famous discovery was the law of gravitation but he also made many experiments on heat and light and he designed new machines for use at the Mint.

red curtains have reflected some sunlight onto it. Red things, like red books, look red because when daylight falls on them they reflect more red than anything else. The light that they send to your eye thus looks red.

Questions

(1) A piece of red material used as in the prism experiment is called a red *filter*. Where have you met this word before? Why is the red glass called this? Is this a good name for the glass when used this way? Does it tell you what the red glass is doing? Which coloured light is like the filtrate mentioned on p. 71?

Fig. 23.26.

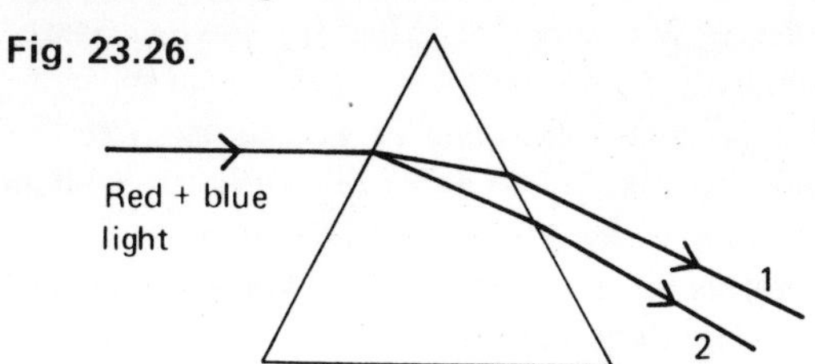

(2) A beam of mixed red and blue light (often called magenta) falls on a prism as shown in Fig. 23.26. Which colour follows path 1 and which follows path 2? Which colour has been deviated most.

(3) Put a red filter in front of your electric torch and shine the light onto red, blue, green objects. Record what they look like. Make a table as below

Colour of filter used	Colour of object in daylight	Colour in filtered light

(4) When you use a filter, some of the light energy gets through as red light. The filter stops the rest of the energy. What kind of energy does it turn into?

Summary

We see an object because light energy from it can get into our eyes. Our brains think that the energy has come straight from the object.

If the rays haven't come straight, then our eyes are deceived and we think the object is somewhere else. We see an image.

When the direction in which the light energy is travelling is altered we say that the light has been deviated.

Light rays can be deviated by reflection at a surface or by refraction at the frontier between two transparent substances.

An ordinary flat mirror gives an image as far behind it as the object is in front.

When light passes through a slab or sheet of glass it comes out in the same direction as it went in.

When light passes through a prism it is deviated; it comes out in a different direction.

Ordinary daylight, or the light from a filament bulb, is a mixture of many different colours.

A colour filter allows some colours to pass through it but stops the others.

Lenses

In your scientific work you have used a hand lens a number of times. You used one to study mixtures and also to show that much heat energy comes to us from the sun by radiation (p. 119). You probably know that lenses are used in cameras, film projectors, microscopes and telescopes, so we ought to find out something about them. If you take a little trouble you can make your own apparatus and if you have bulbs and batteries for them you can do the experiments at home.

Something to do at home

Fig. 23.27.

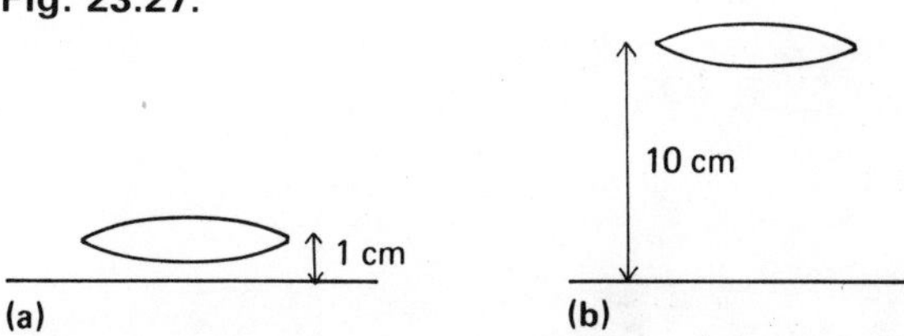

(1) Write your name quite small on a piece of paper. Hold your lens about 1 cm from the paper and look through the lens from a distance of about 50 cm (Fig. 23.27 (*a*)). (This is *not* how you were taught to use the hand lens but never mind that for now!) Does the print look larger? Can you read it?

Lift the lens very gradually a little further from the paper. Does the print now look bigger? Is it really bigger? Are you seeing an image of it? Whereabouts can you read it most easily, in the centre of the lens or round the edges? Raise the lens still further until it is about 10 cm from the paper (Fig. 23.27 (*b*)). You may now have to move your head to see it clearly. What difference do you notice now? Copy the following sentences and complete them:

When the lens is held very close to the print we see an image that is .... and also ....

When the lens is further from the print we see an image that is ....

(2) For your next experiment you will need some home-made equipment. The items needed are shown

Fig. 23.28.

Lens
Screen cut from envelope
Wood
Wood
Wood – with lens mounted

in Fig. 23.28. For a luminous object you can use an ordinary flash-lamp bulb in a plastic holder or arranged as for the electrical experiments in Chapter 19. To hold the lens you can use a cork with a cut in the top or some small support with plasticine on top of it. You will need a little screen. Get your father to save some of those envelopes that have transparent windows to them. He probably gets plenty – most bills come in envelopes like this! Cut a suitable piece and use a slit cork or plasticine to hold it. It is a help to have a second lens available, but you can manage without.

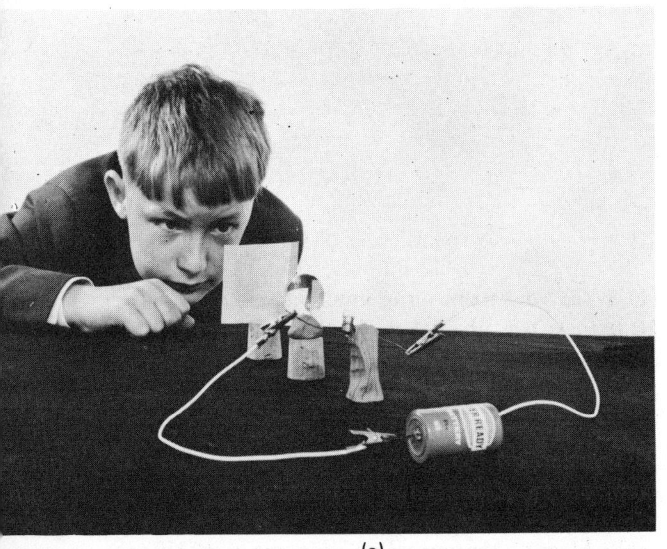

(a)

Fig. 23.29.

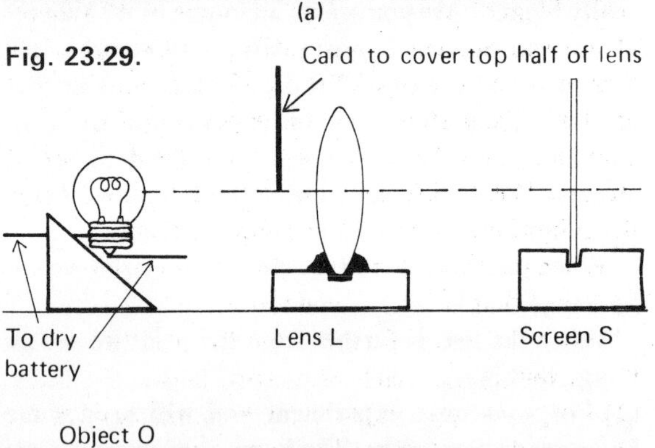

(b)

Arrange your apparatus as in Fig. 23.29 (*b*) and make sure that the three items bulb, lens and screen are on about the same level, as shown by the dotted line.

Put O about 10 cm from L. Move S about and try to find a sharp image on the screen S. This is best done by looking *through* S and moving it until the filament can be seen on the screen as clearly as possible. Is this *really* the filament on the screen? Does the lens really make the image? What happens if you remove the lens? It is helpful if you have a second lens to use as a hand lens while you adjust. You should now be able to see the whole bulb but of course the filament will be brightest. What do you notice specially about this image? Measure LS. When you have measured LS remove S. Can you still see an image of the bulb?

Move O about 2 cm nearer L, and use S to find the image again. Go on doing this until OL is about 6 cm. Then make OL 12 cm and increase OL in 2 cm steps until it is about 20 cm. Make a record of your observations as follows:

Distance OL (cm)	Distance LS (cm)	Image bigger or smaller	Image upside down or right way up

If the image is very small you may find it difficult to see, but do your best. Try also to find an image when OL is 4 cm and 2 cm. If you can't succeed, remove S and just look at the bulb through the lens. What do you see?

The images are made by the lens (how do we know?) and so the lens must deviate the rays to make them. A lens is like a collection of prisms craftily arranged so that all the rays from a single point on one side go through a single point on the other, as in Fig. 23.30.

An image that is bigger than the object we call *magnified,* an image that is smaller we call *diminished.* If the image is the right way up we say it is

Fig. 23.29. An experiment you can do at home with a lens. (*a*) The boy is looking at an image of the bulb on the screen although it does not show up in the photograph. What image can *you* see? (*b*) A diagram of the apparatus.

Fig. 23.30.

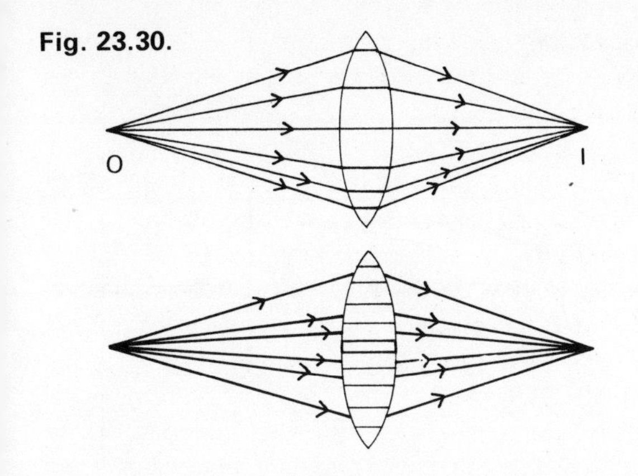

erect and an upside down image we call *inverted.* An image that we can catch on a screen we call a *real* image. Some images can't be caught this way. Look at your own image in a mirror. Would you be able to put a screen behind the mirror and move it until your image was on it? Try it! These other images we call *virtual* images. They are just as useful as the real ones – you use a virtual image every time you make up, or part your hair.

More home experiments

Fig. 23.31.

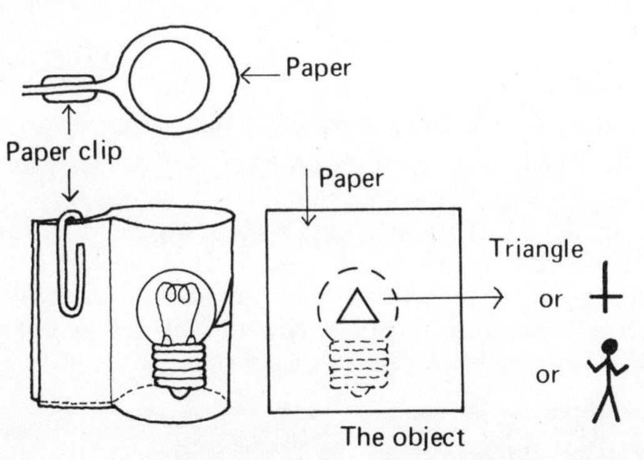

(1) Fit your little bulb with a mask, which you can make from a piece of thin paper as in Fig. 23.31. On the paper draw a little figure or a Δ in ink, fit the paper round the bulb as shown and fix it with a paper clip. Arrange the object, lens and screen as before so as to have on the screen an image rather bigger than the object. Think carefully about how the lens makes an image, and *before* you try the experiment write down your answer to this question: If I cover up the top half of the *lens* with a card, as shown in Fig. 23.29 (*b*), what will happen to the image? Will the top half disappear? or the bottom half? or all of it? or none of it? Then try it. Explain the result if you can.

(2) How far from your lens do you have to hold a

Fig. 23.32.

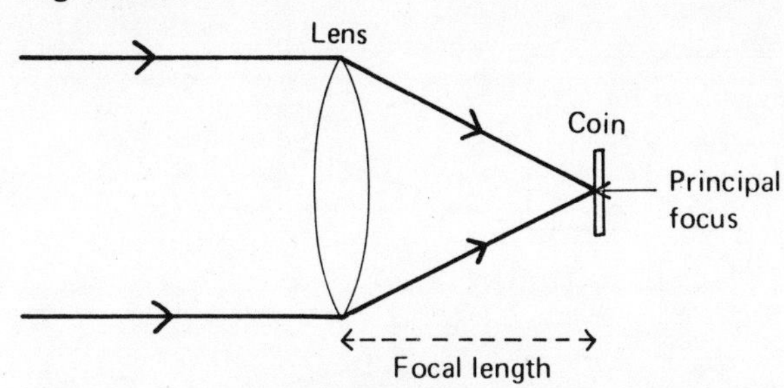

coin to get a sharp image of the sun on your coin? Why would it be foolish to use your screen instead of a coin? Write down this distance. It is called the *focal length* of your lens.

The point at which the lens forms an image of a very distant object like the sun is called its *principal focus*, or just the *focus*, for short. Its distance from the lens is called the focal length (f.l.). If an object is nearer to the lens than its principal focus a virtual image is formed. If it is further from the lens than its principal focus a real image is formed. If scientists know what the focal length of a lens is they can work out where images will be. You will not do this but if you have more than one lens and they have different focal lengths you can do one or two simple experiments with them.

Experiment

Set up your apparatus as in Fig. 23.39. Use the same lens as before and start with an object a good way from the lens, so that the image is small but quite clear. Without moving the holder, take the lens out of it and examine it carefully. Get another lens with the same diameter (meaning on p. 16) but that is thinner in the middle and put it in your holder. The image on S will now be blurred, so move S until the image is sharp again. Which way do you have to move it – towards the lens or away from it?

Fig. 23.33.

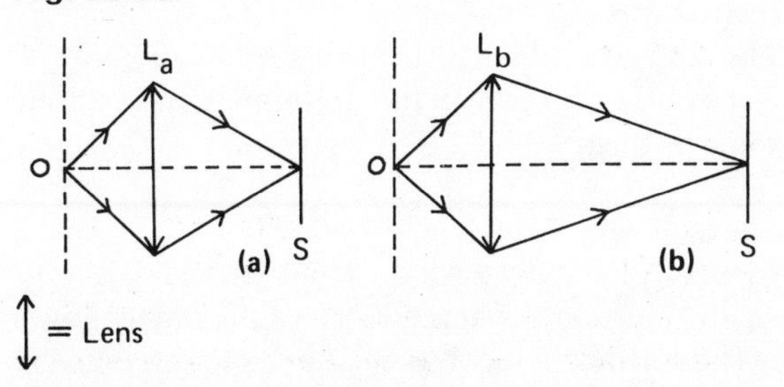

In Fig. 23.33, which of the two lenses, L_a or L_b was fatter in the middle? Their diameters are the same.

Fig. 23.34.

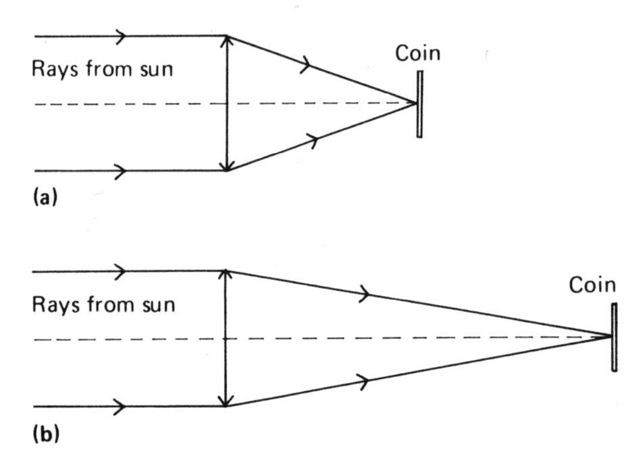

In Fig. 23.34 the lenses have the same diameter. Which of them is fatter in the middle? Which one has the bigger focal length? Which deviates rays of light more? The lens that deviates the rays more is said to have the greater *power*.

Fig. 23.35.

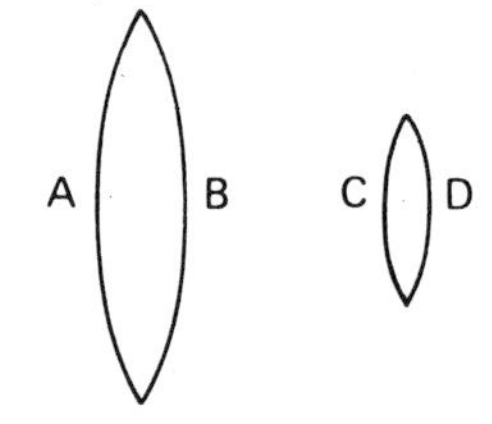

If lenses have different diameters then we can't judge their power just by their thickness. In Fig. 23.35 the surfaces A and C have the same curve on them and so have B and D. These would have the same power but lens AB is thicker because it is bigger. So the thickness of a lens is only a rough guide to how powerful it is. Which lens, AB or CD, would heat something more quickly if you used it as a burning glass?

Instruments

The study of optical instruments is too hard for us just now, but you may like to know a little about some of them.

The hand lens

Figure 23.36 shows why a hand lens makes very small things look bigger. C is a tiny crystal (ours is drawn much too big so that you can see the diagram without a lens). Rays of light from C are deviated by the lens as shown and the light enters your eye as though it had come from C_1. So your eye is deceived and sees a big crystal at C_1.

Fig. 23.36.

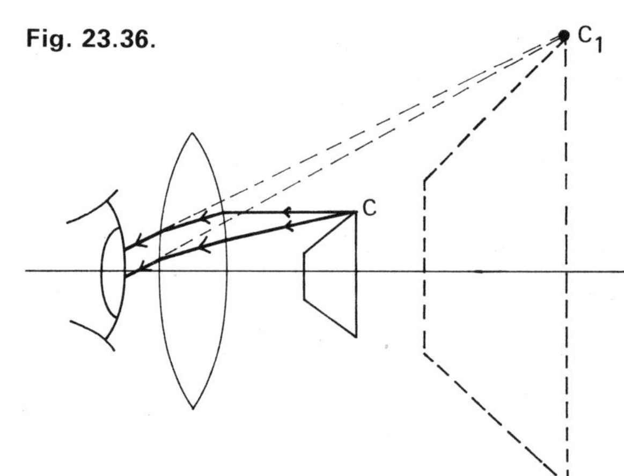

The projector

You made your own little slide projector when you used your lens and screen to give you a magnified image of the figure in front of the little bulb. You can improve on yours if you use a more powerful lamp, like a reading lamp, and a lens with a larger diameter and longer focal length. Full-sized projectors are more complicated, but they all have a lens to make the image, a very bright light to shine on the object (the slide) and a screen. What precaution must you take when you put the slide in the projector? Why?

Cameras

Some of you may own a camera and, if so you know that this also has a lens in it. This lens makes

Fig. 23.37. The projector in a modern cinema. The lamp produces a great deal of heat, so a big tube is provided for ventilation. When the film has been wound round the little wheels the lens will be put in position.

images of objects which are in front of it. The screen is film, a piece of celluloid coated with special substances. One of these is silver bromide. What elements does it contain? Chemical changes take place when light energy reaches these substances. What happens to silver chloride when it is exposed to light? (see p. 105). Silver bromide behaves much the same. After the changes, the film is processed and the picture appears. The camera must be made of material that is completely opaque and the lens is covered by a shutter except when a picture is being taken because otherwise the film would be spoiled. If light could get in all the time it would change all the silver bromide on the film before you got round to taking your snaps.

Does the camera lens make a real image or a virtual one? Is it inverted? Why are your pictures not inverted? Why is the camera made so that the lens can screw in and out? If you had just taken a landscape, would you screw your lens in or out if your next picture was to be a portrait?

The eye

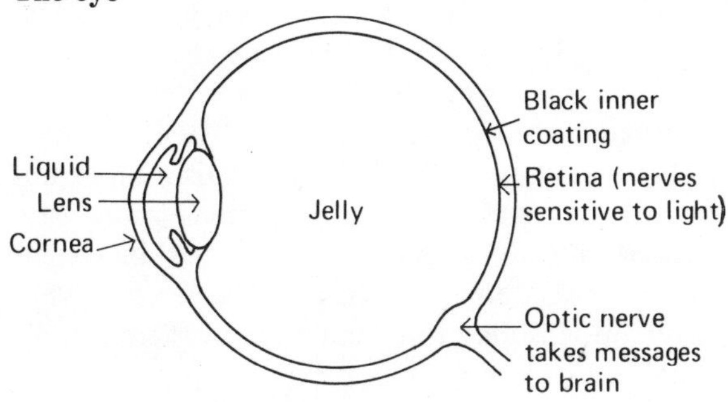

Fig. 23.38. The human eye.

A most wonderful instrument is the eye itself. Fig. 23.38 shows what your right eye would look like if cut through horizontally. Rays of light from the things you look at are refracted at the clear front surface – the *cornea* – and again by the lens so that an image is formed at the back on a sort of screen full of nerves. The screen is called the *retina*. The light affects these nerves and messages go to the brain through the *optic nerve*. The brain tells us what is seen. There is a tough transparent jelly inside the eye to keep it the right shape. If you ask your butcher he may give you a cow's eye or a sheep's eye and you can cut it open and look for the various parts.

The camera is able to take photos of distant scenes and of people nearby because you can move the lens nearer to or further from the film. The eye manages its focussing differently.

Hold your finger in front of a window about 30 cm from your eye and look closely at it. You will find that you can't see distinctly the distant things outside the window while you are inspecting the finger. If you look at the distant scene then the finger looks blurred. When you look out of the window and then at the finger, muscles inside the eye make the lens a bit fatter. This more powerful lens now makes an image on your retina, but you can't see both near and distant things at once, and it takes a second or so for the muscles to work. Try it again.

The coloured part of the eye is called the *iris* and it alters the amount of light energy that gets into your eye.

Something to do at home

Look into a mirror and while you are looking shine an electric torch into your eye. (Don't use a very bright one.) Watch in the mirror what happens to the iris. Why does it happen? Why does it seem dark when you go into a cinema during a performance but less dark after you have been in there a while? Why do your eyes feel a bit strained if you come out from a dark cinema into bright sunlight?

Telescopes

There are many kinds of telescopes, but you can make a simple one quite easily. Get two lenses, one with a long focal length and the other short – your hand lens will do for this. It is best if the first lens has a bigger diameter. Measure their focal lengths as well as you can. (How?)

Fig. 23.39.

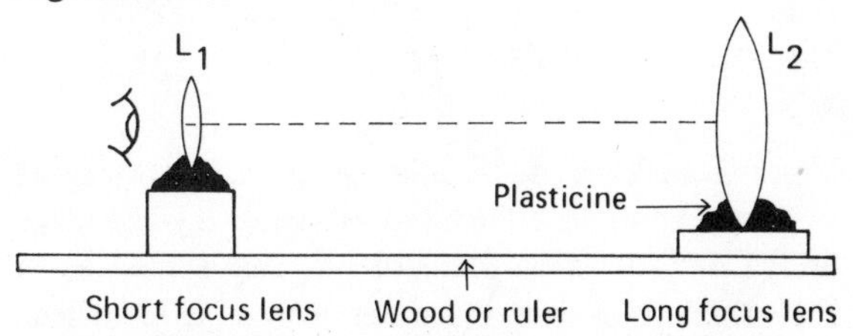

Using plasticine, fix the more powerful lens at one end of a ruler or strip of wood. Fix the other lens so that the distance L_1 L_2 is equal to the two focal lengths added together. Point your telescope at a distant building. If you don't see the image clearly you may have to move one of the lenses a little. Do you see a magnified image? Would the telescope be much use for looking at things? This kind of telescope is much used by astronomers. Why can they manage with it?

Fig. 23.40. The Giant Telescope at Mt Palomar. It is so big that the man using the telescope sits in the end of the tube! He points the telescope where he wants by using electric motors, and he moves with it. This telescope uses a mirror, instead of a lens, to form the image. You can see the stair at the front reflected in the mirror at the end of the tube.

The microscope

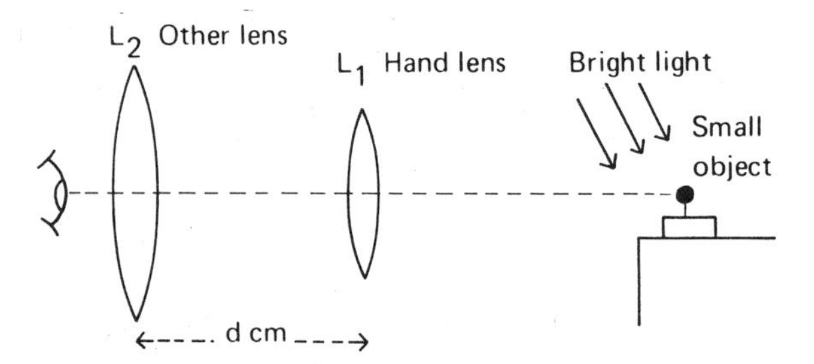

Fig. 23.41. A model microscope.

You can make a model microscope using the same two lenses. First use the hand lens to give you a large image of your lamp filament on your screen as in Fig. 23.29. Now use the other lens as a hand lens and fix it in position so that when you look through it you see the image of the filament on the screen magnified still more. Take away the screen and bulb and measure the distance *d* cm between the two lenses. Fix the lenses as in Fig. 23.41 but with *d* cm between them. If you can fix them in a cardboard tube it will be even better. To examine a small object put it on a cork or something like that and shine a bright light on it. Support your microscope on a book so that it is at the right level with L_1 close to the object. Move the microscope slowly away from the object until you see it magnified. You may find it best to shut out bright daylight and you mustn't let the bright light that lights the object shine into your eyes. Is the image erect or inverted? Why is the bright light necessary?

Fig. 23.42.

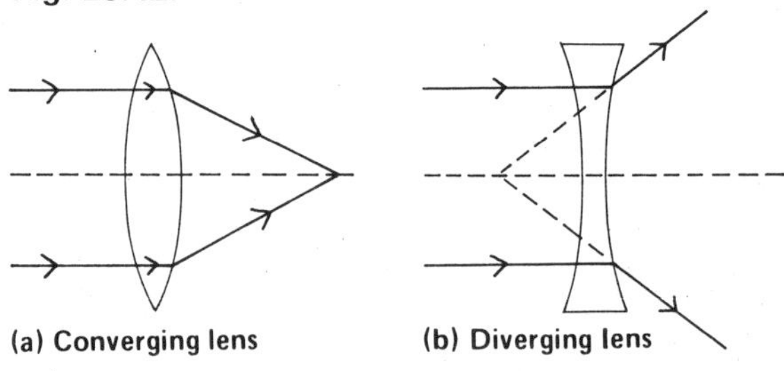

Look at Fig. 23.42. The experiments we have done were all done with lenses that were fatter in the middle than at the outside. They are called *converging* lenses. (Why?) There are also lenses that are thinner in the middle than at the edges – some of you may wear spectacles like this. They are called *diverging* lenses. (Why?) They are also important but are less interesting than the converging kind because they don't magnify or give real images of things. You can't use them as burning glasses. Later on you may learn more about them but we shan't say anything more here.

Summary

A lens deviates the light that falls on it and can make an image of an object put in front of it.

An image that can be caught on a screen is called a real image. If the image can't be caught on a screen we call it a virtual image.

The image of a very distant object is formed at the principal focus of the lens.

As the object comes nearer to the lens the image gets further away and gets bigger.

An object that is closer to the lens than its principal focus gives a magnified virtual image.

Some instruments that use lenses are the camera, the telescope, the microscope and the projector. There is also a little lens inside the eye.

Chapter 24 Gases that Burn

What sort of fuel do you use at home? In one short street, some houses will use gas, others coke, and others oil. Some will only use electricity, but this is made in a power station where the chemical energy of a fuel is changed into electrical energy. Here is an energy diagram:

Fuels burn (in furnace)
→heat energy (in boiler)
→shape energy (of steam)
→movement energy (steam in turbine)
→electrical energy.

The fuels boil water in the boiler and this takes up a lot of space. The steam then moves along pipes into a turbine which acts like a windmill. Its parts spin round at a very high speed, and they are connected to the generator which gives us our electricity.

What gas do the fuels combine with? We will now look more carefully at one or two fuels, finding out where they come from, how they are made, and what their properties are.

Gas

Have you ever seen those huge tanks which are used to store gas? In Fig. 24.1 (*b*) you can see a diagram of one. Gas is made in the gas works and pipes carry it to the gas holder for storage. How do we collect gases in the laboratory? The gas holder works in the same kind of way – gas is collected by displacement of water. As the gas goes into the holder the dome rises. Will the weights over the pulleys help this? Have you ever heard your mother

Fig. 24.1. A gas holder. (*a*) You may have seen one like this in your town. When gas has been made it is stored ready for use. Pipes will carry the fuel from the holder to your home. What prevents the gas from escaping? How is the large area of metal prevented from rusting? Why do men have to use the ladders? Can you see any difference in design between the empty holder on the right and the one in the centre? (*b*) shows a diagram of a gas holder.

(a)

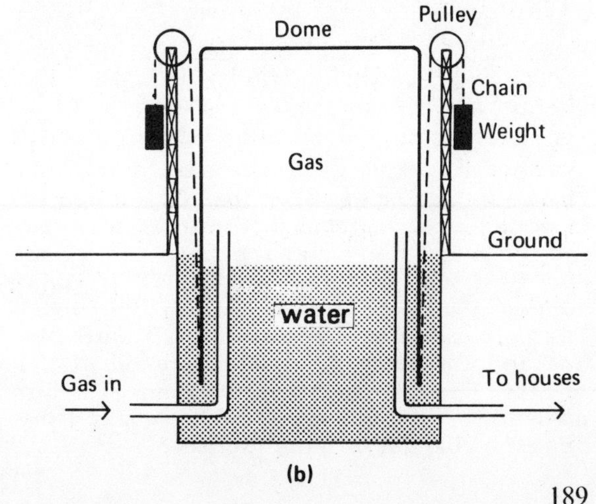

(b)

say that the gas pressure is low? She means that gas is coming out slowly. How could you change the holder to make it come out faster?

How gas is made from coal

Not so very long ago all our gas was made from coal. Here is an experiment which shows how this is done.

You will need some coal which has been broken into small pieces. The apparatus is shown in Fig. 24.2. Half fill the test tube with coal and connect it to the one which is half full of water. Heat the coal with a Bunsen burner using a small flame at first, and keeping the flame moving so that all the coal gets hot. From time to time try to light the gas coming out of the jet.

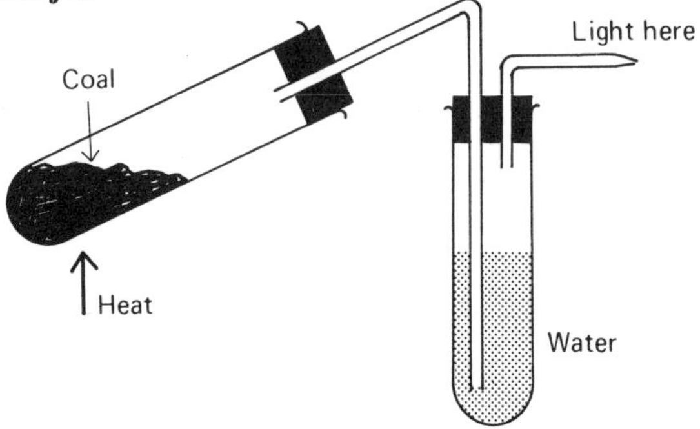

Fig. 24.2. Making gas from coal.

Did you get a gas which burns? What was the flame like – was it clean or sooty? Why? Did anything collect on the water? Do you think that air can reach the hot coal? What would happen if it did? Was anything left from the coal?

Gas works make their gas in this kind of way. The gas is heated in iron tubes called retorts, and no air is allowed in so the gas cannot burn. You can see the tops of some retorts in Fig. 24.3. When it is very hot the coal decomposes and it gives off gas and tar. Coke is left behind in the retort. The tar is collected in big tanks full of water – what does the cold water do? Gas bubbles on and is collected in the gas holder.

Fig. 24.3. Making gas from coal. Coal is heated in sealed steel vessels called retorts. You can see the metal funnels at the top so that coal may be tipped into the retorts. What would happen if the pieces were not the right size? The light-coloured pipes on the left carry the gas and tar away and coke remains in the retorts. The two men are working in a hot noisy place – why is it hot? Why noisy? How does the coal get into the funnels?

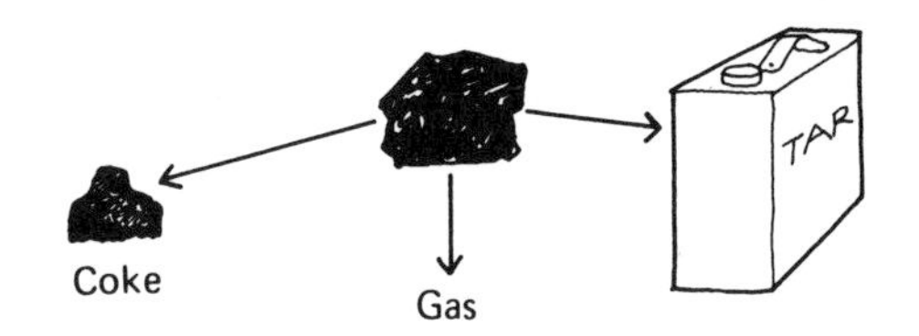

Fig. 24.4. Coal, a most important chemical.

In Fig. 24.4 you can see some of the uses for coal. Tar is a mixture of many different chemicals and your teacher may let you look at some of them. Here is a list:
benzene, phenol, cresol, pyridine, naphthalene, anthracene. Look at the pictures of their molecules in Fig. 24.5. Count the atoms in each of them.

Notice that the molecules in tar are made of rings of atoms. There are single rings, and there are some molecules where rings are stuck together. Later on you will see that the molecules in oil are rather different.

The tar is taken to special factories where all these chemicals are separated. What ways do you know for separating mixtures? All these chemicals have different boiling points – which way would you try first?

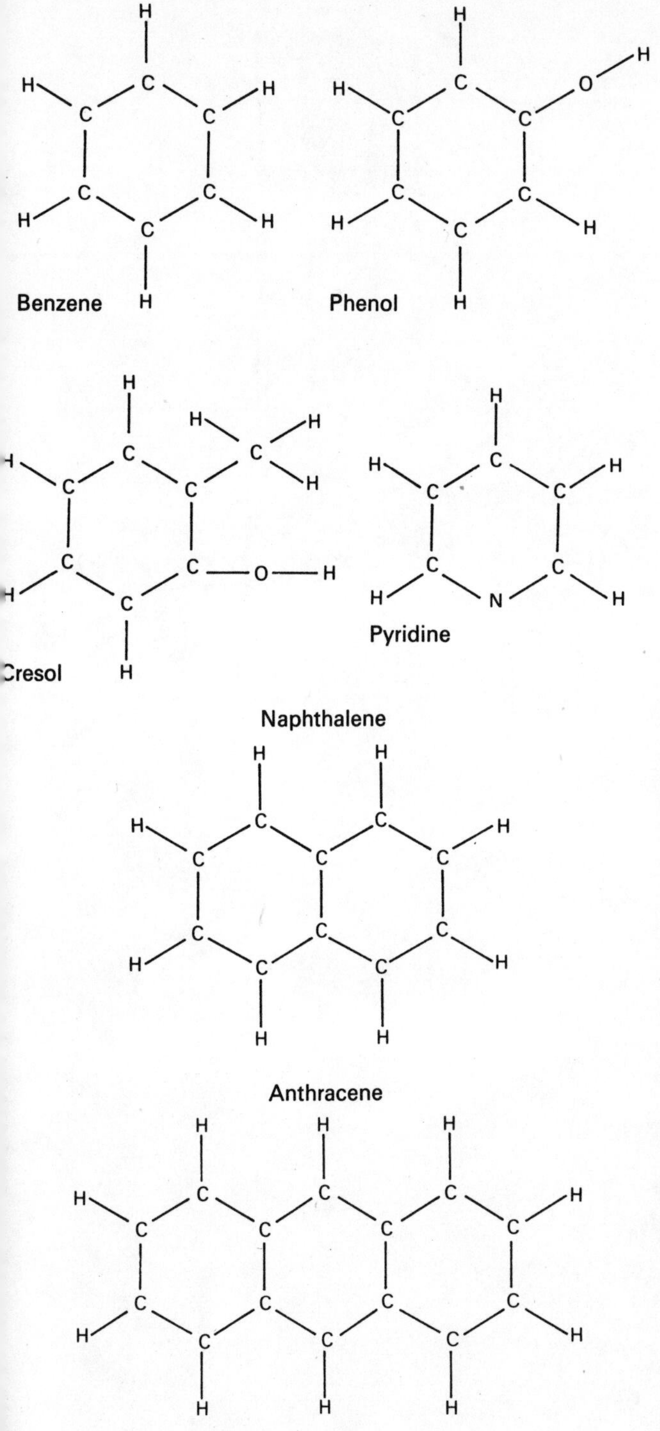

Benzene
Phenol
Cresol
Pyridine
Naphthalene
Anthracene

Fig. 24.5. Some of the molecules in tar.

Distilling tar

First of all pour 2 cm³ of tar into the special tube shown in Fig. 24.6. Be careful to avoid getting it on your clothes or hands. Pour it into the tube using a filter funnel, but take it in turns to use the same funnel – this will save on the washing up. Put the cork and thermometer (reading up to 300° C) into the tube and then clamp the tube at Bunsen height. Put a dry tube ready to collect liquids which distil over. Now heat the tar with a low flame. When the temperature is 140° C change the collecting tube for another. When the temperature is 220° C stop heating.

Thermometer (300°C)
Rubber sleeve
Glass tube
Tar
Heat
Distillate

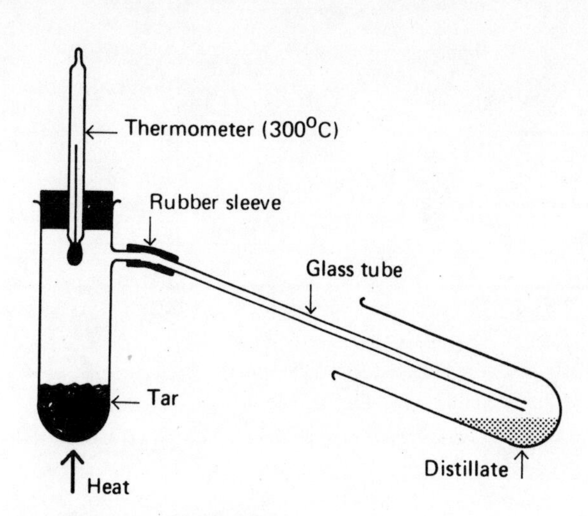

Fig. 24.6. Distilling tar.

Natural gas

When engineers want to get oil from under the ground they drill deep holes through the rock until they reach it. Sometimes they drill in the sea, and in Fig. 24.7 you can see a picture of a 'rig' working in the North Sea. The drill is lowered to the sea bed and then drilling begins.

Sometimes the oil has a lot of gas dissolved in it and this is called *natural gas*. It is a very good fuel. In this country we buy some natural gas from North Africa. How do you think we might carry the gas all this way? It would be an enormous job to lay a pipeline! What happens when we cool a gas? In Fig. 24.8 you can see a ship specially built to carry *liquid* natural gas. Its cargo is very cold indeed and it must not be allowed to warm up on the long journey. What colour would you paint the liquid holders? How would you make them? How would you change the liquid back to a gas when you arrived in port? The molecules of natural gas contain hydrogen and carbon atoms. Unlike wax molecules, those in the gas are very small and they only have one or two carbons. What will the products of combustion be when natural gas burns?

We need such a lot of fuel gas that scientists have worked out a way of making it from oil itself. The oil is heated so that it becomes a gas and this is then

Fig. 24.7. Some of the gas we use at home comes from under the North Sea. This rig is drilling into the rock on the sea bed and with luck it will strike gas. Can you see the drill? What is the platform marked 'H' for? Why does the rig have four rather than three legs? When the drill strikes gas the engineers must plug the hole until they can set a pipe over it. Then the gas can be piped ashore.

Fig. 24.8. Methane Progress was specially built to carry liquid natural gas. Why was it called 'Methane' Progress? How is the natural gas liquefied? Do you think that the ship is fully loaded?

mixed with steam. When the mixture is passed over a special catalyst there is a chemical change and fuel gas is made. Scientists call this way of making gas *steam reforming*.

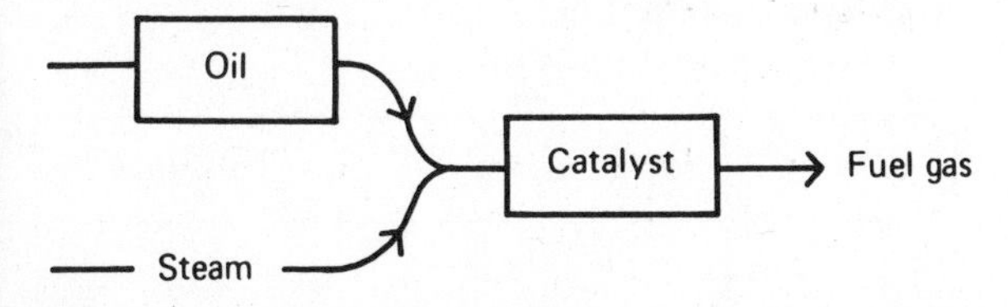

Fig. 24.9. Steam reforming.

Here are some of the molecules in oil. In steam reforming they are broken up so that we get hydrogen, carbon monoxide, carbon dioxide, and other things. Notice that the molecules in oil are like chains. What are the molecules in tar like?

```
   H  H  H  H  H
H—C—C—C—C—C—H          pentane
   H  H  H  H  H

   H  H  H  H  H  H
H—C—C—C—C—C—C—H        hexane
   H  H  H  H  H  H

                H
              H—C—H
   H  H  H  H  H  |  H
H—C—C—C—C—C—C—C—H      2 methyl
   H  H  H  H  H  H  H     septane
```

Fig. 24.10. Molecules in oil.

Fuel gas is a mixture

The gas we burn at home or in the laboratory is not a pure chemical compound, but a mixture of different gases. You will have heard of some of them but others may be new to you.

Gas	Molecule
Hydrogen	H—H
Carbon Monoxide	C—O
Methane	H H—C—H H
Ethylene	H H C – C H H
Acetylene	H—C—C—H

How many atoms are there in a molecule of methane? What kind of atoms are they? Gas does not have fixed amounts of these ingredients – this depends on how it is made. But there is so much hydrogen in our gas that it is worth studying it on its own. Hydrogen is an element of course. What does this mean?

How to make hydrogen in the laboratory

Some time ago you passed an electric current

through some liquids, and you saw that the electricity decomposed the liquids to give simpler chemicals. Name two which do this. What chemicals did they give?

It ought to be an easy job to make hydrogen because such a lot of compounds have it. You have heard of some of them already – give the names of a few. Here is a list of some chemicals which have hydrogen, and you may have thought of many others – can you write their molecules?

Name	Molecule
Water	H—O—H
Hydrogen chloride	H—Cl
Sulphuric acid	H—O O S H—O O
Nitric acid	O H—O—N O
Sodium hydroxide	Na—O—H

It may be that heat energy or electrical energy can enable us to get hydrogen from these compounds, but we will try something new.

Suppose your dog has a stick between his teeth and will not put it down? How would you encourage him to drop it? How about giving him something he would like to have even more than a stick? Why not offer a bone so that he drops the stick?

Now let us tackle the problem of making atoms 'drop' hydrogen. We must offer them something instead. I wonder what the atoms will prefer?

What we must do is to take a chemical with hydrogen in it and then put in different elements to see if their atoms will take the place of hydrogen. Hydrochloric acid has simple molecules and we will use this first. Here is a list of the elements you may know: magnesium, zinc, iron, copper, sulphur, carbon.

See if these can do the job.

Be careful with hydrochloric acid. If you get it on your hands, wash them. You will need a test tube rack and clean test tubes. Quarter fill each with the dilute acid. Get ready for the elements by writing their names on small pieces of paper. Your teacher will then give them out.

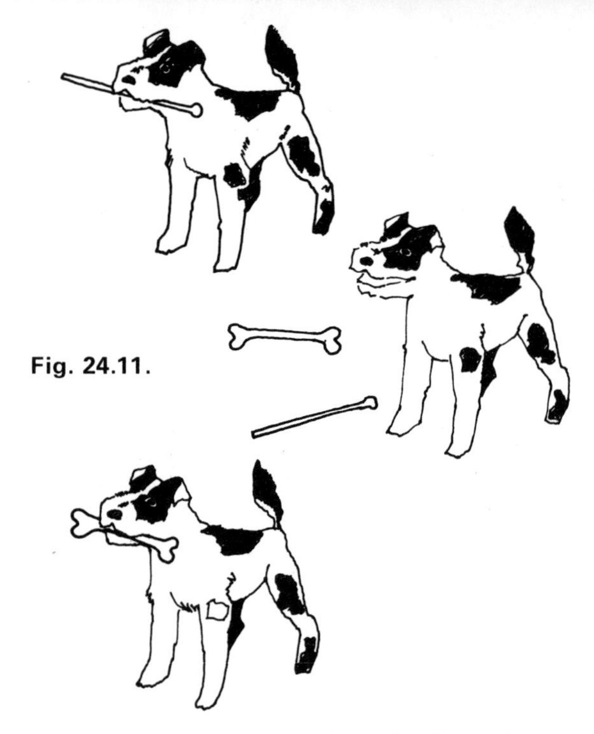

Fig. 24.11.

Put a little magnesium into the first tube, zinc into the second and so on. Watch the tubes carefully. If you think hydrogen is bubbling off, test for it with a lighted splint. What happens? Write a table in your note-book and fill it in.

Element	Observation	Any hydrogen

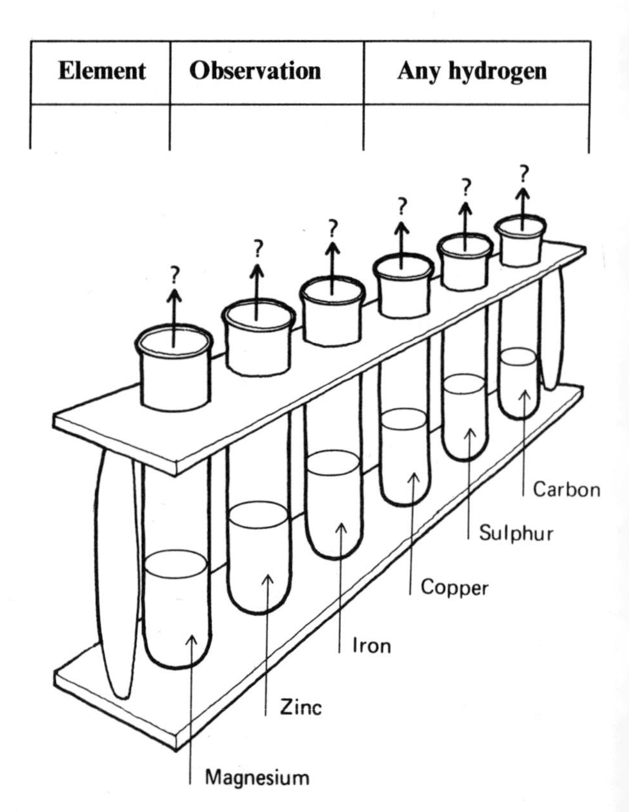

Fig. 24.12. Trying elements on hydrochloric acid.

If you have time see what some other elements will do when they are put with hydrochloric acid. But do ask before you try them.

When one kind of atom pushes out another we call it a *displacement reaction*. You have seen how copper atoms are pushed out by zinc atoms; this is another displacement reaction.

Do you think that you could displace the hydrogen atoms from sulphuric acid using other atoms? Repeat the experiment you have just done but this time put dilute sulphuric acid in the tubes. Put your observations into a table again.

You now know that we can get hydrogen in many different ways. Which do you think is the easiest? We must now go on to collect some hydrogen so that we can learn about its properties.

Make a sketch of the apparatus you used to make oxygen from hydrogen peroxide. We can use this same apparatus to make and collect some hydrogen. Put some zinc pieces into the boiling tube and add dilute sulphuric acid using the dropper. Collect the gas by making it displace water. As one tube fills up change it for another, but do not take the tubes out of the water until you are ready to use the gas. It will escape from the tube very quickly if you do – why?

Now you can write about the properties of the gas. Answer the following questions in your note-book: Has it a colour? (we have a special name for things which have no colour). Does it smell? (We have a special name.) Put a lighted splint to the gas. What happens?

The dangers of hydrogen/air mixtures

We use a lighted splint as a test for hydrogen. But did you notice that the gas does different things when we do the test? Some tubes 'pop' loudly, some burn quietly with a blue flame. What makes this difference? What is hydrogen combining with when we do the test? Have you ever heard a similar 'pop' when you light the gas fire or the gas water heater? If you are one of those people who turns on the gas and then looks for the lighter you will get a loud 'pop' every time! If you have a lighter ready then you will not get it. Hydrogen burns quietly in air, but if you have a mixture of hydrogen and air the burning is very fast indeed – we call it an *explosion*.

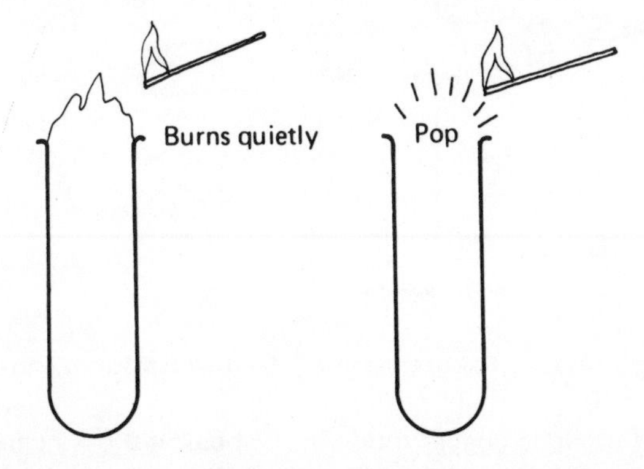

Fig. 24.13. Hydrogen burns, hydrogen/air explodes.

Fig. 24.14. When gas is mixed with air there is a risk of explosions. A lady living in this upstairs flat left a gas tap on without lighting the cooker, and a short time later the gas-filled room exploded. Notice the violence of the explosion.

Fuel gas is like hydrogen and it does the same thing. You may have seen pictures of houses or flats blown down by gas explosions. What has happened is that gas and air have mixed in a room and then a light has made the two explode with great force. You may like to see just how dangerous a mixture can be.

The tin in Fig. 24.15 has a hole in its lid and another in the bottom. Hold the tin so that the bottom hole is against the gas tap, and turn the gas on full so that gas streams into the tin. After a few seconds turn the gas off and put the tin on a tripod with its lid at the top. Light the gas at the top hole with a long taper. Now

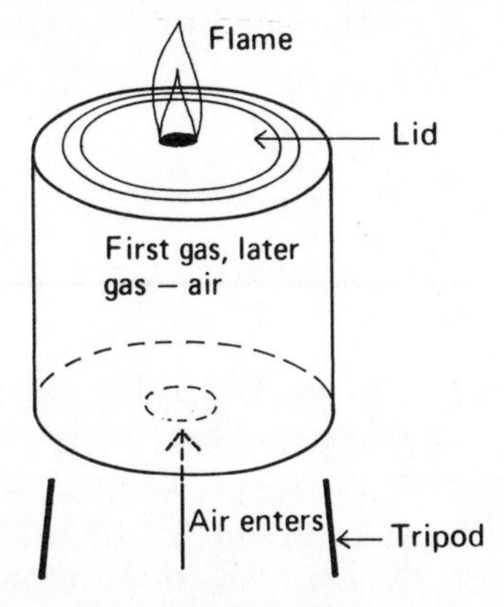

Fig. 24.15. Gas/air explosions.

stand back and watch the flame carefully. You may have to wait a minute or two.

As gas burns at the top what will happen at the bottom hole? Why was the flame yellow at first and blue later? What happened to the lid? Did it get any energy? What kinds? Where did the energy come from? Think what may happen if a cooker is full of a gas/air mixture. This is why we must have a light ready.

There is a lot of hydrogen in 'gas', but when hydrogen is on its own it is even more dangerous. It will be better if some of its properties are shown to you – let your teacher take the risks!

Hydrogen is the lightest element we know having a density of 0.00009 g/cm^3. What does this mean? Have you seen it used for flying balloons? Ask your teacher to fill one with the gas and attach a label with your name and address. Let it go. One of mine was sent back from Poland.

What is made when hydrogen burns?

What chemicals are made when a candle burns? What are the products of combustion of 'gas'? When hydrogen burns it will give a compound whose molecules have oxygen and hydrogen atoms in them. What compound is this? Now we will see if we can collect the compound.

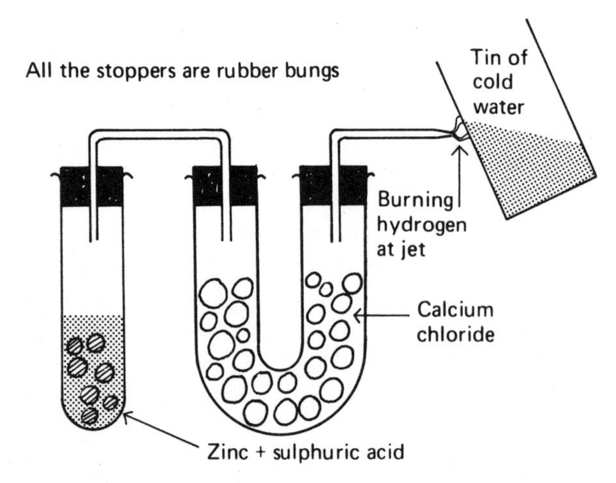

Fig. 24.16. Burning hydrogen and collecting the product.

Set up the apparatus like that in Fig 24.16. *Do not light the jet.* **When hydrogen has been coming out of the jet for two minutes, cover all the apparatus with a heavy towel so that only the jet and the tin are showing. Light the jet. Look at the tin where the flame touches it.**

Do you agree that the liquid on the tin looks like water? What was the calcium chloride for? The hydrogen is made from zinc and dilute acid and there is a lot of frothing. What will leave the boiling tube mixed with the hydrogen? If the calcium chloride was not there could you be sure that the water came from the burning hydrogen?

When hydrogen burns in air it combines with the oxygen to give water. As a chemist you will know that this compound ought to be called hydrogen oxide because it is made up of oxygen and hydrogen atoms. Draw a molecule of water in your note-book. Hydrogen combines with oxygen – what do we call a reaction like this? What has happened to the hydrogen?

Can hydrogen take oxygen from chemicals?

We have seen that hydrogen will combine with oxygen gas to form water. But will it combine with oxygen if this is already part of a compound? In the next experiment we will see if it can do this.

Write down the names of ten compounds which contain oxygen. For this next experiment we will use copper oxide, and instead of hydrogen we will use ordinary 'gas'. Remember that this has a lot of hydrogen in it and that it is safer (there is no air with it). Make the apparatus shown in Fig. 24.17. Here are a few tips to help you:

(1) Clamp the combustion tube high enough to get a Bunsen under it. Also clamp it at one end so that you can heat the middle of it.

(2) Make it slope slightly. Why?

(3) Do not put the porcelain boat in the tube yet and do not turn the gas on.

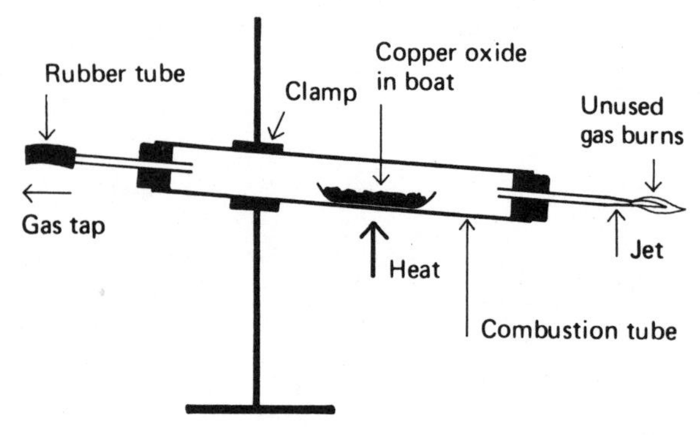

Fig. 24.17. Taking oxygen from copper oxide.

Put some copper oxide in the boat and take care that none is sticking on its outside. Take out the cork and jet and slide the boat into the tube – a pencil is

useful for pushing it far enough. Now put the cork and jet back. Turn on the gas so that it flows over the copper oxide, and after a few seconds light the jet. (This is a good way of getting rid of unused gas.) The jet flame should be about 3 cm long. Light a Bunsen burner to give a medium sized clean flame and then put it under the boat. When nothing else seems to happen take the Bunsen away and let the boat cool down with gas still flowing over it. Take the boat out and examine the contents.

Did hydrogen take oxygen from copper oxide?

What colour did the black powder go? Would you expect to see water in the tube if oxygen was removed by the hydrogen? Did you see any water? Why was the tube sloping? What happens to hot copper in air? Why did we cool the boat down with the gas still flowing?

You can now try the experiment with other oxides. Someone may like to use magnesium oxide, some iron oxide, some lead oxide, some mercury oxide. Share your results and make a table in your note-book.

Name of oxide	What we saw	Did the hydrogen take oxygen?

This is a kind of displacement reaction. Which element is the 'dog' and which 'the bone'?

Hydrogen takes oxygen from a few compounds; it works with copper oxide, lead oxide, and mercury oxide. But it does not work with magnesium oxide, calcium oxide and others. When a compound has oxygen taken from it we say that it has been *reduced.* 'Hydrogen reduces copper oxide'.

copper oxide + hydrogen ⟶ copper + water.

We call hydrogen a *reducing agent* because it is able to take oxygen from some compounds.

When hydrogen reduces copper oxide what adds to the hydrogen? What has happened to the hydrogen? You will see that in any reaction where a compound is reduced, something else must be oxidised at the same time.

Copper oxide loses oxygen	Hydrogen gains oxygen
Copper oxide is reduced	Hydrogen is oxidised

Can we find a more powerful reducer than hydrogen?
Magnesium combines with oxygen gas very violently. Will it be a good reducer? We cannot pass magnesium over a metal oxide because it is a solid, so we will try a different way. We will also give it a hard job to do – hydrogen could not take the oxygen from zinc oxide, but can magnesium?

Mix a teaspoonful of magnesium powder and zinc oxide on a piece of paper. Pour the mixture onto a piece of asbestos so that it makes a neat pile. Put a piece of magnesium ribbon into the pile so that it acts as a fuse (Fig. 24.18). Now light the ribbon holding the Bunsen at arms length.

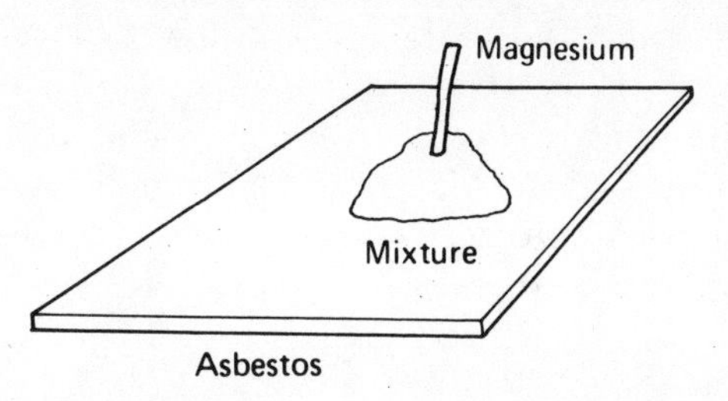

Fig. 24.18. Magnesium, a powerful reducer.

Did the reaction happen? Complete the following:

zinc oxide + magnesium → ------ + ------

Is magnesium more powerful than hydrogen at reducing things? Can you design an experiment to see if zinc will reduce copper oxide? Think one out and your teacher may let you try it. The reactions inside fireworks are oxidations and reductions – be very careful if you are thinking of doing some oxidations at home. Remember the rules:

Ask your teacher before you try,
use very small quantities,
never mix solids with a stirrer which may make a spark,
do not bend over a mixture,
wait a long time before you think that a mixture will not work.

The uses of hydrogen

Some chemical works use hundreds of tonnes of hydrogen every week because it goes into the making of some very important substances. They do not make their hydrogen from zinc and acid – this is far too expensive – but instead they get it from fuel gases. Sometimes they get it from water – how? Some of the important uses of hydrogen are shown on the next page.

Fig. 24.19. Making ammonia. It is very difficult to combine hydrogen and nitrogen to make ammonia gas. In this tower, the two gases are passed over a catalyst which speeds up the reaction so that some ammonia is obtained quickly. The tower can make about 1000 tonnes of ammonia a day. Many chemists spent a lot of time inventing the catalyst for this tower – but ammonia is a most important chemical. How will the chemists get the nitrogen and hydrogen? How high is the tower? Look at the rungs of the ladder.

Making ammonia. You may have heard of this compound at home – find out if your mother has ever bought any, and find out why. It is a compound containing the elements hydrogen and nitrogen. Where do we get the nitrogen from? The molecule of ammonia is shown below:

$$\begin{array}{c} \mathrm{H} \\ | \\ \mathrm{H{-}N{-}H} \end{array}$$

How many atoms are in it? How many of each? Ammonia is a very useful cleaner, and it is also used to make nitric acid and fertilisers. It is made by mixing the hydrogen and nitrogen at very high pressure and temperature and passing the mixture over a special catalyst.

Making margarine. To look at margarine you wouldn't think that hydrogen had a part in it – but it has. To see how it is made look at Fig. 24.20. Vegetable oil and hydrogen are made to react by a catalyst (an element called nickel), and margarine is the result.

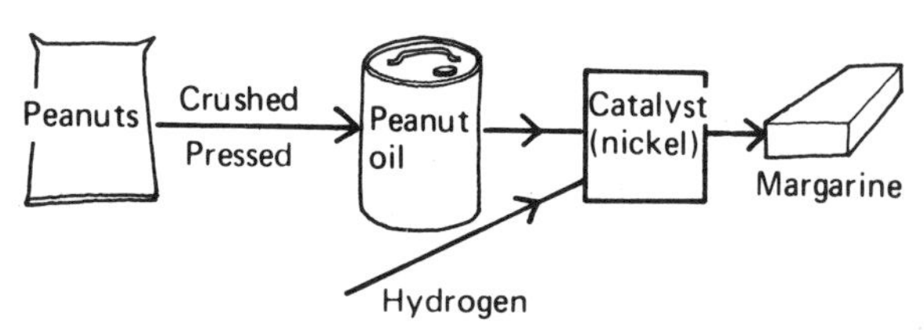

Fig. 24.20. Making margarine.

Summary

Fuel gas is made by heating coal without air so that it cannot burn. We also get coke and tar in this way and these are very useful too. Sometimes we get fuel gas from oil wells; either as natural gas, or by making the oil react with steam. Fuel gas is a mixture and it has a lot of hydrogen in it; it also has some methane and carbon monoxide.

We can make hydrogen in the laboratory by mixing acid with some metals – zinc and sulphuric acid are very good. Hydrogen has a lot of interesting properties. It is colourless, odourless, tasteless, and it burns in air to give water. It makes an explosive mixture with air. It is a reducer, taking oxygen out of some compounds. It is a most useful gas, going to make margarine, ammonia and nitric acid.

Questions

(1) What is coke, and where does it come from?
(2) What is tar, and where does it come from?
(3) Write notes on two ways of making fuel gas.
(4) Give the names of three compounds which are in fuel gas and draw their molecules.

(5) Catalysts are used a lot in industry. Write about three uses of them.

(6) Draw a diagram of the apparatus you would use to prepare some hydrogen. Label it.

(7) Write down the names of ten compounds which contain hydrogen, and draw their molecules.

(8) Give two examples of,

a displacement reaction,
an oxidation reaction,
a reduction reaction.

(9) How could you find out whether carbon dioxide was a reducer? Draw a labelled diagram as part of your answer.

(10) How could you find out whether iron was a reducer or not?

(11) Could carbon dioxide be an oxidiser? How would you find out?

(12) Robert Boyle was one of the first scientists to study hydrogen. Write about him and include the following information: when he was born and when he died, where he lived, what experiments he did.

Chapter 25 **Solid Fuel**

Have you ever heard your father complain about having to clean out the fire grate or the stove? Or do you ever have to have the chimney sweep come to your house? Ashes, cinders and soot are made when we use solid fuels to warm our houses or do our cooking.

There are many kinds of solid fuels:

Coal. This is dug out of the ground. Coal miners work deep underground, using picks and shovels, and sometimes high explosives and power saws to loosen it from the rocks above and below it. The coal is then carried to the surface in special lifts called cages. What kind of energy does it gain? Where does this come from? It is then sorted into pieces of different sizes and finally washed.

When coal burns it makes a lot of smoke. Because of this many people think that we ought not to burn it; after all, the smoke falls onto the streets and makes them dirty, and it can make some people very ill indeed. In *smokeless zones* people are not allowed to burn coal, and they have to use something else.

Anthracite. This is a special kind of coal, and not many mines produce it. When it burns it gives out a lot of heat and it leaves very little ash behind. Many people use it in stoves for warming the kitchen and making hot water. They like having no ash to clean up – but they don't like paying the bill for this very expensive solid fuel. We call anthracite a smokeless fuel because it does not give off very much smoke.

Coke. How is coke made? You may have done an experiment to make fuel gas from coal; can you draw the apparatus you used and give the names of the three things made in the experiment? Good coke burns to give hardly any ash and smoke. But it is not easy to light and it needs a very good supply of air for it to burn properly. Do you have a gas poker at home? What for? Does your father 'riddle' the stove every night and every morning? Why?

Carbon: another element

Name a few compounds which have carbon in them. If you can give six you are doing well; if you can give pictures of their molecules you are doing very well indeed!

You will all have carbon dioxide on your list. How do you test for this gas? How would you make carbon change to carbon dioxide? Here is another way.

Mix one teaspoonful of carbon powder and one of copper oxide on a piece of paper. Put the mixture into a clean dry test tube and fit this with a delivery tube like the one shown in Fig. 25.2. Half fill another tube with limewater and dip the delivery tube into it.

Fig. 25.1. Here you can see a miner using a machine to tear coal from the coal face. What kind of energy makes the machine work? Can you see the metal beams which hold up the roof? How are the blades of the cutter made very hard? Notice that the miner is wearing a helmet fitted with an electric lamp.

Clamp the apparatus as you can see, and keep the clamp well away from the mixture. Now heat the mixture gently, using a low flame. As soon as you stop heating the mixture, take the delivery tube off – why?

(1) What happened to the mixture?

(2) What happened to the limewater?

(3) Can we use this as a test for carbon?

(4) If we do, ought you to do the experiment with copper oxide on its own? Why?

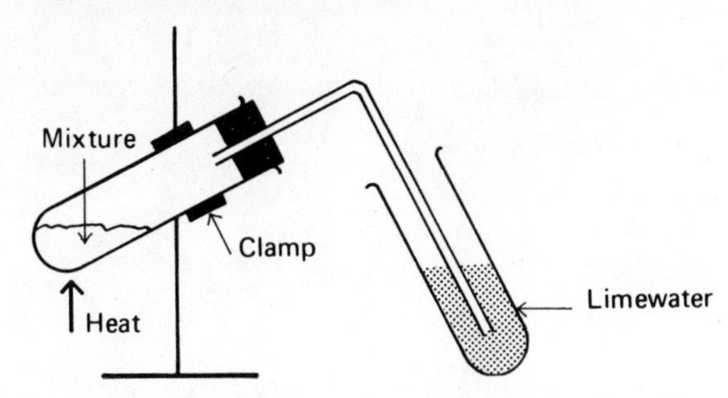

Fig. 25.2. Changing carbon to carbon dioxide.

Now using powdered coal, coke, anthracite – and any other solid fuel you may have, find out whether they have carbon in them. Some of you can do one whilst others are doing another.

(*a*) **Do not have your test tube more than $\frac{1}{4}$ full of mixture.**

(*b*) **Stop heating as soon as you have an answer to your question.**

(*c*) **Scrape out the tubes using a glass rod – do not use water and do not put waste into the sink (it will block the drains).**

Solid fuels are mainly carbon. You can see what they are made of, and how much heat they give when they burn, from the table.

Carbon in solid fuels

Name	Carbon	Hydrogen	Oxygen	Joules of heat for each gramme burned
Wood	$\frac{1}{2}$	$\frac{1}{10}$	$\frac{4}{10}$	16000
Peat	$\frac{6}{10}$	$\frac{1}{10}$	$\frac{3}{10}$	20000
Coal	$\frac{9}{10}$	$\frac{1}{20}$	$\frac{1}{20}$	30000
Anthracite	$\frac{95}{100}$	$\frac{3}{100}$	$\frac{2}{100}$	32000
Coke	Can be all carbon	—	—	32000

How good are these fuels?

Write down what a 'joule' is. You will see that the fuels with a lot of carbon in them give a lot of heat when they burn. When you have a bonfire next, notice the colour inside it. If it is going well it will be a nice red colour. But if you look at a coke fire which is going well, you will see that it is white hot. This is because the coke is giving more heat and the fire is hotter.

Work out the rise in temperature when 1 g of coke heats 100 g of water. Now work out what 1 g of wood will do.

In Fig. 25.3 you can see three wagons, and each of them can give the same amount of heat energy.

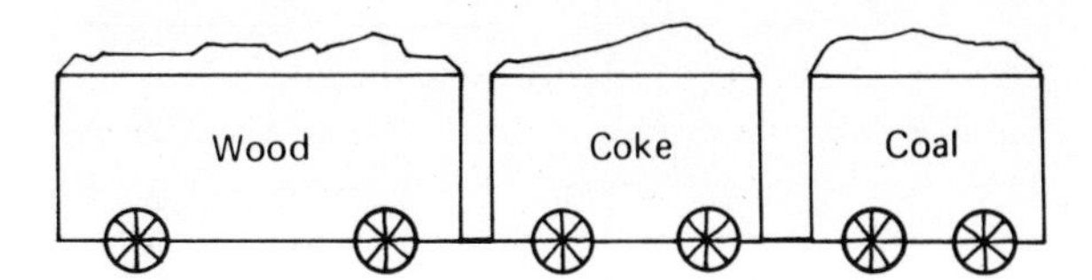

Fig. 25.3. Heat from fuels.

Which fuel gives most heat for each gram? Why is the coke wagon bigger than the coal wagon? Why is the wood wagon so big?

Carbon – the element with many disguises

You can see below a picture of some diamonds. You all know what a piece of coke looks like. The quite remarkable thing about these two substances is that they are the same element – carbon. You may ask how we know this. The answer is that they both

Fig. 25.4. When diamonds are found they look like this. All of them are made of carbon but the dark ones have a little impurity in them. To make them sparkle they must be cut so as to have flat faces and the skilful diamond cutter does this without wasting much material. You can see how they look after cutting on p. 9. Which city is famous for its skilled diamond cutters?

burn in exactly the same way; they give carbon dioxide only. This means that they are made up of carbon atoms only. You may have heard of graphite – this is carbon too.

How can we have different kinds of carbon? How can the same atoms go to make hard brilliant diamonds or dull coke? In Fig. 25.5 you can see a picture of two walls, but one is stronger than the other. Which is the stronger wall? Is your house built like (*a*) or (*b*). The walls are so different because the bricks go together in a different way. It is the same with diamond and coke. The carbon atoms are arranged differently.

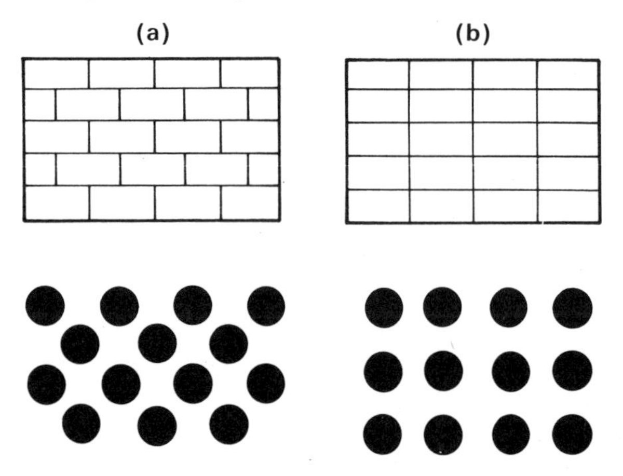

Fig. 25.5. Different arrangements of atoms.

When an element can be different things we say that it shows *allotropy*, and each kind of the element is called an *allotrope*. We can say that diamond and coke are allotropes of carbon.

Some properties of carbon

(1) You already know that carbon has allotropes. Check up that you know what this means.

(2) You also know that carbon burns in oxygen – what does it give?

(3) Does carbon conduct electricity? See p. 156.

(4) Diamond has a density of 3·5 g per cm³. Graphite has a density of 2·3 g per cm³. How could you check up on these figures? See p. 19.

(5) Diamond has a specific heat capacity of 0·52 J/g°C (see Chapter 27). Graphite has a specific heat capacity of 0·64 J/g°C. How could you measure these? See p. 213.

Now we will learn some new properties. What happens when iron, zinc or copper are mixed with hydrochloric acid, sulphuric acid and nitric acid? Do you think carbon will combine with these acids?

(a)

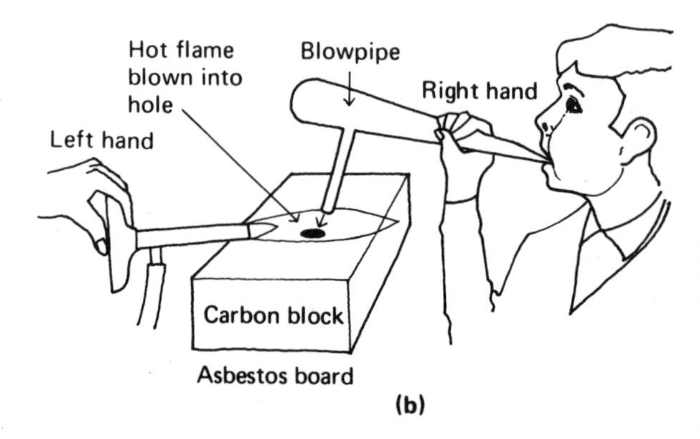

(b)

Fig. 25.6. Carbon the reducer. The charcoal block (carbon) is able to reduce copper oxide and lead oxide. (*a*) Notice the neat hole in the block and the way the flame is directed to make the hole very hot. Why does the boy hold the block using tongs? Why would it be wrong for him to put the block on the bench as soon as he has finished heating it? See if you can blow through your mouth and breathe through your nose at the same time (this boy can). (*b*) is a diagram of the apparatus.

Put three test tubes in a rack and pour 3 cm of hydrochloric acid into one, a similar amount of sulphuric acid into the second and nitric acid into a third. Add a spatula of carbon powder to each. What happens? Would it help to warm them? Carbon does not react with these acids.

Is carbon a reducer?

What happened when hot carbon was put with oxygen gas? What was made? Can carbon join up with oxygen if this is already chemically combined? We found out that one element could do this – what is it?

For this experiment you will need a charcoal block and a blowpipe. The charcoal block is made from carbon powder which has been pressed together. First make a small hole in the block by drilling it with a pen-knife. Use a spatula to fill the hole with lead oxide and press it down. Put the block on an asbestos board. Now arrange the block near a Bunsen flame and use your blowpipe to make the lead oxide very hot. Keep blowing till the lead oxide is a red hot liquid (you will have to practise to get this right). Let the block cool down on the asbestos square – ***never*** **put it on the bench – and then dig out the chemicals inside the block using your pen-knife. Is there any lead? Is the carbon a reducer? What may happen if you use copper oxide instead of lead oxide? Try it – but thoroughly clean the block first.**

Carbon is a reducer and it can take oxygen from some compounds:

Copper oxide + carbon ⟶ copper + carbon dioxide
Lead oxide + carbon ⟶ lead + carbon dioxide

Some uses for carbon

We all know that carbon in the form of diamonds makes lovely jewelry. But diamonds have a more important use because of their hardness. If you want to cut wood you use a chisel made of steel because steel is very much harder than wood. But suppose we want to cut steel to make a chisel! Well, we use cutting machines made of even harder metal. You will now ask how we cut the blades for a cutting machine and the answer is that we use tools tipped with diamonds to do this. Industrial diamonds are not as big or as clear as the ones in jewelry – but they are still very expensive.

We make iron from its ore by putting the ore and coke and limestone into a blast furnace. When air is blown into the furnace the inside gets very hot (1500° C) and the coke *reduces* the iron ore (iron oxide).

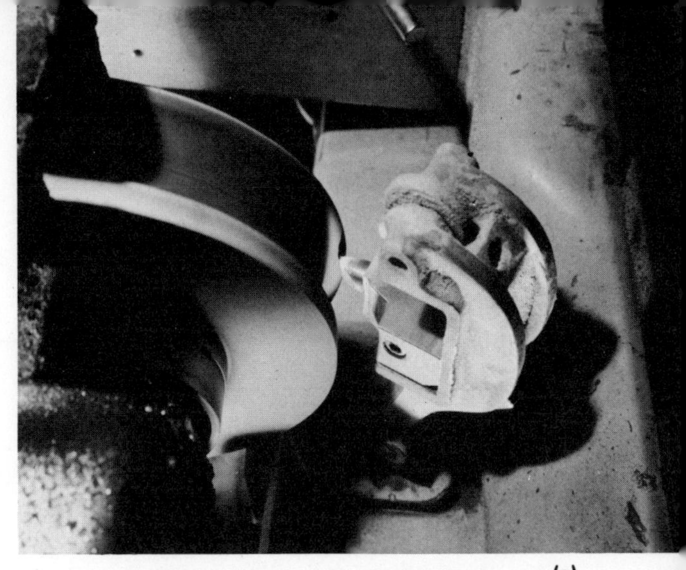

(a)

Fig. 25.7. Diamonds are used to tip tools which must cut through very hard materials. (*a*) The circular saw is used to cut paving stones and its blades are lined with diamonds. How would you trim pieces off a grinding wheel – the kind of wheel which is used to sharpen knives? You can see a tool working on such a wheel and the tool has a diamond tip. (*b*) In the installation of a de-icing system on the Chiswick flyover, power supply cables were fed from the central reserve through channels sawn in the granite kerbstones by a diamond bladed, self-propelled saw. This proved much quicker and more economical than taking the kerbs back to precision machines at the stone-yards.

(b)

(1) How do we get coke?
(2) Why can't we get iron from iron oxide using a blowpipe and charcoal block?
(3) Iron melts at 1400° C. What will happen to the iron in a blast furnace?
(4) What gas will come out of X in Fig. 25.8?

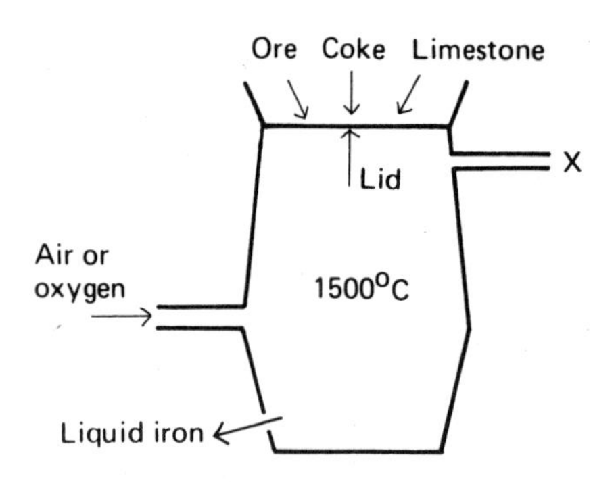

Fig. 25.8. A blast furnace.

If you examine a piece of graphite you will see that it feels slippery. Some machines are lubricated with graphite.

Making hydrogen for industry

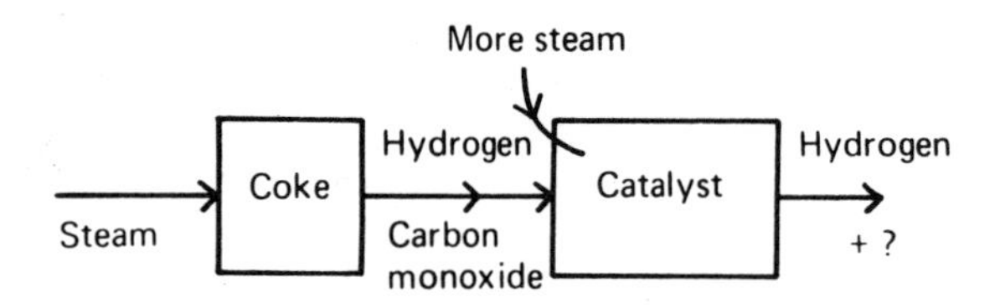

Fig. 25.9. Making hydrogen.

Coke is used for making hydrogen gas in industry. Give some uses for the hydrogen. What might coke reduce so that hydrogen is left? Some big chemical works make a gas called *water gas*. Steam is passed over red hot coke and a reaction happens:

Carbon + water (hydrogen oxide) ⟶ hydrogen + carbon monoxide

What has been reduced? What has been oxidised?

To get the hydrogen carbon monoxide must be taken away and this is done by mixing it with more steam and then passing it over a catalyst. This makes another oxidation – reduction reaction work:

Carbon monoxide + hydrogen oxide ⟶ hydrogen + ?

What is the other substance formed? Which chemical is the reducer? How can we remove the other chemical so that only hydrogen is left?

Summary

Carbon has several forms called allotropes; in these the atoms are arranged differently. Diamond is very hard and is useful for making very hard tools. Coke is useful for making hydrogen and iron and it is also used as a fuel. Carbon does not react with acids. It is a reducing agent and can take oxygen from some compounds.

Questions

(1) Draw the molecules of carbon monoxide and carbon dioxide.
(2) Draw a diagram of a blast furnace and explain what the air does, what the coke does, and what happens to the iron ore.
(3) Find out where diamonds come from.
(4) 12 g of diamond burns to give 44 g of carbon dioxide and *nothing else*. How much carbon dioxide will 3 g of coke give?
(5) When we use a charcoal block for reducing copper oxide the block loses weight (ignore drilling the hole). Why is this? Write five lines.
(6) Complete the following equation and label the oxidiser and the reducer:

Carbon dioxide + carbon ⟶

(7) Sulphur has allotropes. Find out their names and how they are made. Your teacher may let you try when you have found out. Look in the Chemistry Section in your library.

Chapter 26 Making Interesting Compounds

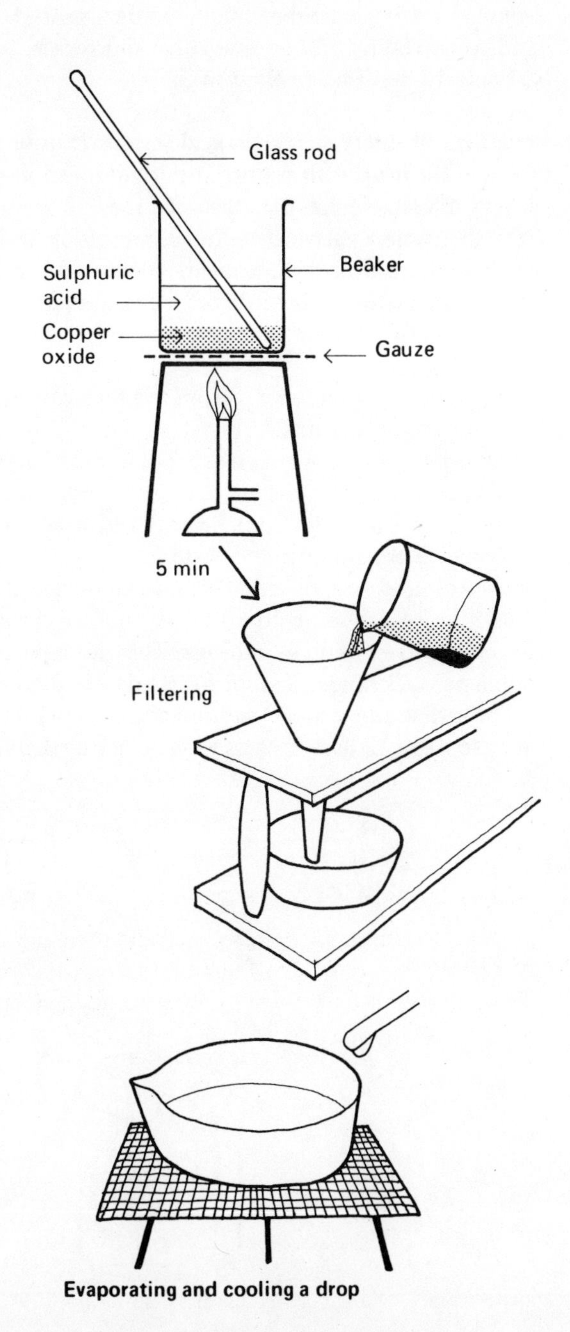

Fig. 26.1. Making copper sulphate.

A Blue vitriol (copper sulphate)

Put 50 cm³ of dilute sulphuric acid in a beaker and warm it on a tripod and gauze. Add to it two heaped spatulas of copper oxide. Stir the mixture with a glass rod and keep it hot for five minutes. If all the copper oxide dissolves, add some more. Get ready to filter your mixture and put an evaporating dish to collect the filtrate. Now filter your mixture. Examine the residue and make a note of what it looks like – you can then put it in the waste bin.

Put the blue filtrate on a tripod and gauze and boil it with a medium sized flame. *Do not boil it dry.* **From time to time dip a glass rod into the liquid then take it out with a drop of the liquid on it. Watch the drop cool. If it gives a solid when it cools, stop heating the dish and put it in a cool place. Leave it there until the next class. Scrape out the crystals onto a piece of newspaper so that the moisture is soaked up. Then put them on a watch glass and dry them on a radiator or in an oven.**

What colour was the residue? What was it? When you made copper sulphate was the change physical or chemical?

B Green vitriol (ferrous sulphate)

Put a few iron nails in a beaker and cover them with dilute sulphuric acid. Cover the mouth of the beaker with a piece of paper and stand it on the window sill overnight. Decant (pour off) the liquid into an evaporating dish leaving the nails behind. Evaporate the liquid – using a glass rod as shown opposite to test when to stop heating. Leave the hot liquid to cool. Handle the crystals as you did for copper sulphate.

Did the reaction give a gas? What was it? Why cover the beaker? Why is it important to avoid overheating the chemical you have made?

C Table salt (sodium chloride)
Common table salt is mined or taken from the sea. But you will find this laboratory preparation interesting. It makes use of a coloured substance called litmus which can show whether water has acid in it. The litmus has been used to dye a piece of paper.

Put a piece of litmus paper into a test tube of pure water. Now fill another tube with pure water and add one drop of dilute hydrochloric acid to it. Add a piece of litmus paper. Do this again, but use any other kind of acid.

We call litmus an *indicator* because it shows (indicates) whether a liquid is acid.

Half fill a test tube with dilute hydrochloric acid and pour it into a beaker. Add a few drops of sodium hydroxide solution and stir the mixture. Dip a piece of litmus paper in to see if there is still some acid there. Take the paper out. Add more sodium hydroxide. Stir again, test with litmus. Go on until all the acid is used up. Does the mixture get warm?

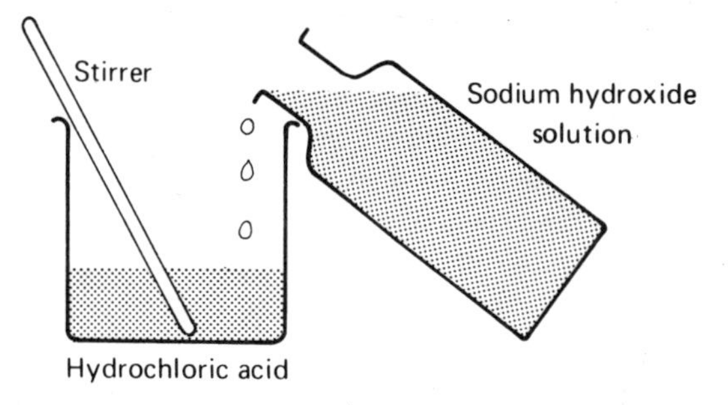

Fig. 26.2. Making table salt.

Pour the mixture into an evaporating dish and evaporate the water until crystals form. When the dish is cool – taste the crystals.

Did you see anything happening when you mixed the chemicals? Did the mixture get warm? What does this mean? Where is the heat coming from? Is the change physical or chemical? Write an energy diagram for this change.

D Invisible ink (cobalt chloride solution)
Pour 50 cm³ of dilute hydrochloric acid into a beaker and add two spatulas of cobalt oxide. Warm the mixture for 10 minutes on a tripod and gauze and stir it with a glass rod. If possible, leave the mixture overnight. Filter and discard the residue. Divide the filtrate into two parts. Evaporate one and try to get the crystals but do not boil it dry – see p. 205. Use the other part as ink. Write on a piece of paper with it and then warm the paper. What kind of change takes place when you warm the paper?

Some kinds of chemical compounds
Acids
You have often used acids in your experiments. Write down the names of three of them. Acids are so important that we must now learn a lot more about them. Remember that acids must be handled carefully. They may be dangerous – particularly the concentrated ones.

Pour 1 cm of dilute sulphuric acid into a clean test tube and fill it up with water. Dip a glass rod into the very dilute acid and then taste the rod.

Do the same experiment with dilute nitric acid and then dilute hydrochloric acid. How would you describe the tastes of these three? Some people say they are 'sharp' or 'sour' or 'tart'.

As you have taken chemicals from the shelves, you may have seen other acids.

Acetic acid – an acid in vinegar. Dilute some and taste it.

Citric acid – an acid from lemons and oranges. Dissolve a crystal of it and taste it.

Tartaric acid – dissolve a crystal and taste it.

Acids may be solids or liquids and some of them may be found in everyday things. They all have a sour taste. Of course, testing for acids by tasting may be rather dangerous – or unpleasant to say the least – so we must use other ways of detecting them.

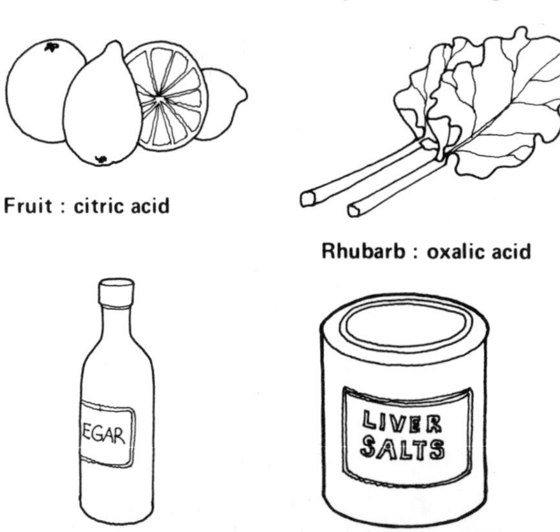

Fig. 26.3. Everyday acids.

What happens when we put litmus paper into hydrochloric acid? What do we call litmus? Check up that it will detect all the acids you know.

Make dilute solutions of the acids and put some blue litmus paper into each one. What happens to the colour of the litmus paper?

Something to do at home

Your teacher will give you some litmus paper so that you can find acids at home. Make a solution of soda and when you want to turn the paper blue again dip it in this and then rinse it with water.

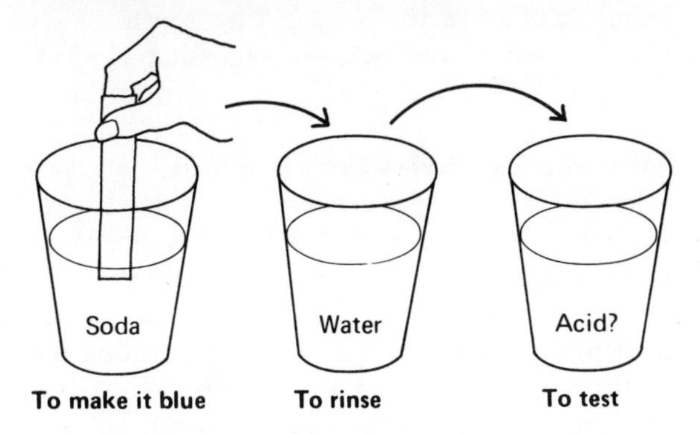

Fig. 26.4. Testing for acids.

Stir the following with water and then test for acid: liver salts, soda water, bleach (very dilute or else you will bleach the litmus), vinegar, salt, fertiliser, baking powder, lemon juice, garden soil. Make a list of substances which have acids in them.

We know that acids are not just chemicals on the shelf in the laboratory; they are common everywhere. There are lots of them and some are very strong and harmful, while others are weak and quite harmless. They have a 'sharp' taste and they turn litmus red. They also fizz and make carbon dioxide gas if we put soda on them.

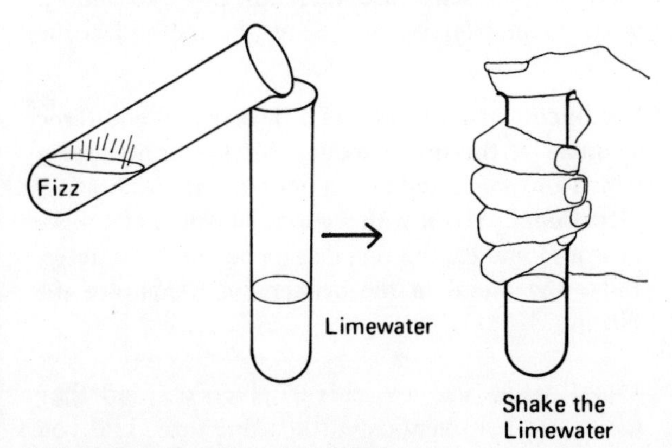

Fig. 26.5. Testing the 'fizz'.

More about molecules

Here are pictures of molecules of some of the acids we have been using.

Hydrochloric acid (hydrogen chloride)

```
H—Cl
```

Sulphuric acid

```
H—O   O
    \ /
     S
    / \
H—O   O
```

Nitric acid

```
          O
         /
H—O—N
         \
          O
```

Oxalic acid

```
H—O   O
    \ /
     C
     |
     C
    / \
H—O   O
```

Acetic acid

```
H—O   O
    \ /
     C
     |
  H—C—H
     |
     H
```

What atom do they all have? Look back to p. 83. All acids have hydrogen in them – but don't be confused – all substances with hydrogen are not acids. Think of a substance with hydrogen in it which is *not* an acid. Let us put it another way. Without hydrogen in it a chemical cannot be acid; if it has hydrogen it may be an acid.

Salts

When we add iron or magnesium to acid we get hydrogen and a liquid, and with soda acids give carbon dioxide and a liquid. We must find out more about these liquids.

Half fill a tube with dilute hydrochloric acid. Add a spatula of soda so that it fizzes. Is there still some acid in the tube? How would you find out? When you have thought of a test on the liquid, try it out. If there is some acid there add some more soda, wait for fizzing to stop and then test again. When all the acid is used up, what is there in the liquid? Or is it just

water? Filter to remove any soda left over and then evaporate the liquid (p. 205). But remember – do not boil the evaporating dish dry. Describe what you see in the dish. Damp your finger, touch the solid so that a little sticks to it and taste it.

Work out an experiment to find out if anything is left in the liquid when dilute sulphuric acid has been used up by magnesium. When you have had it checked, carry it out. Taste it. It is safe to taste these two compounds, but you should never taste any chemical without asking your teacher – some are very poisonous.

When acids are used up by metals or soda, a gas is given off and we are left with a solution from which we can get solids by evaporation. These solids can dissolve in water and they make crystals. They also have a bitter taste. We call these solids *salts*. You will see that scientists can make a lot of different salts because there are a lot of different acids and many different metals. The 'salt' we use at table, common salt, is called by scientists 'sodium chloride'. What substances did you use to make it? How many different salts can you make from zinc, magnesium, hydrochloric acid, sulphuric acid and nitric acid?

Naming salts

In one family I know there are two girls and three boys. The girls are called Edith Smith and Susan Smith, the boys John Smith, Ian Smith and Timothy Smith. Because they are in the same family they all have the same surname, but they have different Christian names of course. Now that you know the names of the children you can give the surname of the father, because they are named after him. Salts are given names in the same kind of way; each acid is the parent of a family of salts because they are made from it. Here are parts of three families of salts:

Family 1	Family 2	Family 3
zinc nitrate	zinc sulphate	zinc chloride
magnesium nitrate	magnesium sulphate	magnesium chloride
copper nitrate	iron sulphate	
	copper sulphate	

First of all look at the second name – you will see that it is the same within each family. This is because all in the family have the same parent acid and so they take their name from it:

'nitrates' are made by using up nitric acid,

'sulphates' are made by using up sulphuric acid, and

'chlorides' are made by using up hydrochloric acid.

You will have guessed what the first name tells us. Which metal has used up nitric acid to make zinc nitrate? Which to make copper nitrate? Which chemicals could you mix to make magnesium sulphate? Zinc chloride? Write the following in your note-books, but fill in the blanks:

Zinc + ... → zinc chloride + hydrogen.
Iron + sulphuric acid → .... + hydrogen.
.... + hydrochloric acid → magnesium chloride + hydrogen.

Some other chemicals which use up acids

What chemicals did you use to make blue copper sulphate? to make cobalt chloride? See p. 205. In these interesting preparations we used up acid, but we did not do it with a metal or soda. Which elements does copper oxide have? Let us find out if other metal oxides can use up acids.

Here is a list of metal oxides	Here is a list of acids
magnesium oxide	sulphuric acid
iron oxide	nitric acid
zinc oxide	hydrochloric acid
lead oxide	
calcium oxide	

It would take far too long for you to try all these experiments, so the best thing to do is to try one reaction each. One of you can see if magnesium oxide uses up sulphuric acid, another can try magnesium oxide and nitric acid, a third magnesium oxide and hydrochloric acid. Another three can use iron oxide with the different acids, etc. How many experiments are there to do? No matter what chemicals you are using, the method is the same.

Put 50 cm³ of the acid into a beaker and add three spatulas of the metal oxide. Put the beaker on a tripod and gauze and heat it until it is almost boiling. Occasionally stir it with a glass rod (why not a piece of iron?) and let the reaction go on for 10 minutes. Filter the liquid in the beaker and evaporate the filtrate.

Metal oxides use up acids to give salts, but they make another chemical at the same time. Did you see what it was? Was it hydrogen or oxygen? Could

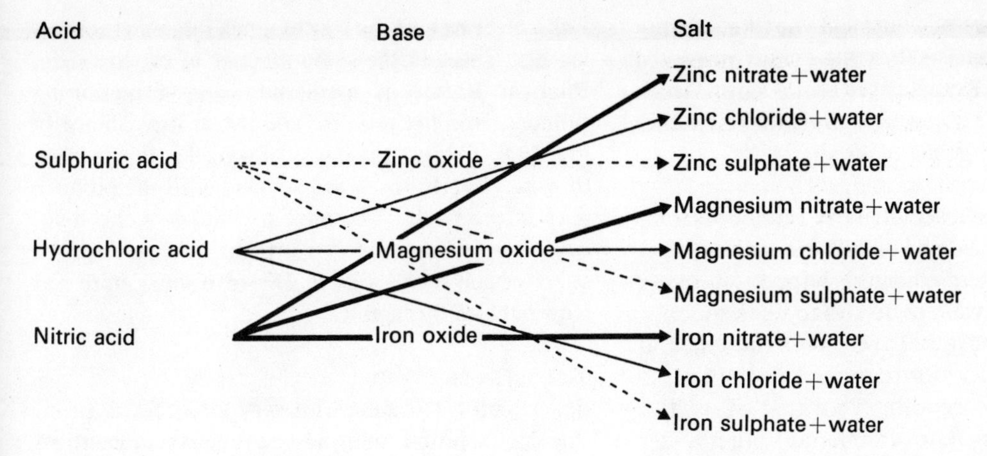

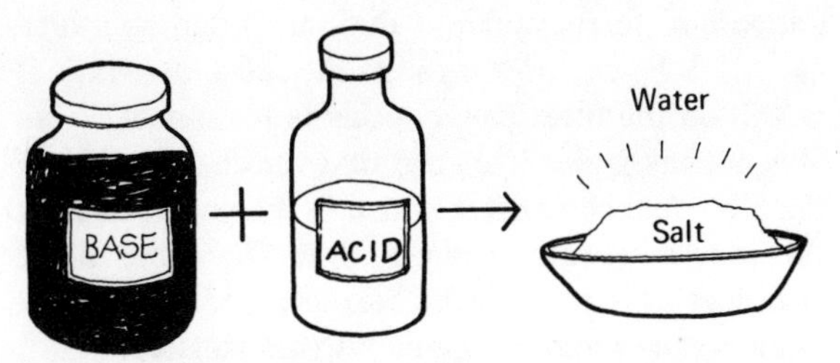

Fig. 26.6. Acids and bases.

it have been carbon dioxide? When magnesium uses up acid:

magnesium + acid ⟶ salt + hydrogen.

When magnesium oxide is used:

magnesium oxide + acid ⟶ salt + hydrogen oxide.

What is the usual name for hydrogen oxide?

Long and difficult experiments show us that metal oxides with acids give a salt and water. It is very hard to show that we get some water because the acids we use have it in them – we use *dilute* acid remember. It would take us too long to do these experiments ourselves so we will take the word of other scientists.

Chemical compounds which contain a metal and oxygen are called *bases* and they use up acids to make a salt and water.

Solutions which use up acids

Most of the substances which use up acids do not dissolve in water. Name three of them. However, a few do dissolve. You have used one – what is it? How did you make table salt? In this preparation we use a solution to use up some acid. What was the acid and what was the solution?

Take a piece of the sodium hydroxide in a test tube and add a little water to it. Feel the tube and then put some red litmus paper into it.

Sodium hydroxide solution – we sometimes call it *caustic soda* – can use up acids. Let us try to find others. Here is a list of solutions you may find in the laboratory:

ammonia, potassium hydroxide (caustic potash), lime water, alcohol and water, acetone and water.

Take a little of each in separate test tubes and put in some red litmus paper. Are any of them like caustic soda? Rub a little of each solution between your fingers. Wash your hands afterwards.

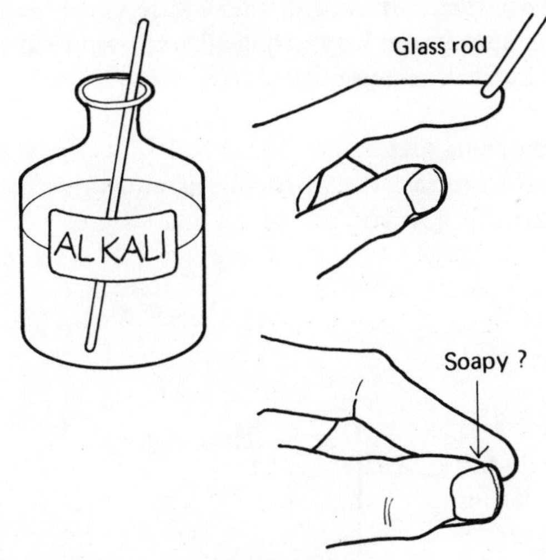

Fig. 26.7. Testing for alkalis.

Ammonia, caustic potash, limewater, are called *alkalis*. They feel soapy, turn litmus blue, and use up acids to make salts.

Put $\frac{1}{2}$ test tube of caustic potash (potassium hydroxide) solution in an evaporating dish. Float a piece

of litmus in it and then add any acid you like, stirring the alkali all the time. When the litmus turns red what is the liquid? Evaporate to make sure. Which salt have you made? If you have time, make another salt from a different acid and alkali.

Occasionally we have accidents in the laboratory, and concentrated acids or alkalis are spilt and get onto hands or clothes. When this happens it is very important to know what to do and to work quickly. First of all we must get well away from the spill and then pour plenty of water onto it. This makes the corrosive liquid more dilute so that it is less dangerous. If the acid is on your hand – put it under the tap. If it is on your clothes, sponge them with a very wet cloth – but try not to spread the acid too much. Your teacher will then do the rest. If it is an acid spill he may sponge it with very dilute soda or ammonia. Why? If it is an alkali he may use dilute acetic acid – why? Dilute acids and alkalis are not nearly so dangerous, but you have learned to be careful with them. Clothes are very expensive and if you splash them you may spoil them. Why is the scientist on p. 90 wearing a smock?

Some of the things we use at home have dangerous acids or alkalis in them. The liquid in a car battery is very strong acid and you may have seen what happens when it splashes on metal under the bonnet. Never put your head over and near to the hole in a battery when you put distilled water in it. Some oven cleaners are very strong alkalis – wear rubber gloves when you use them.

Make some soap

Alkalis are very useful chemicals and this experiment will show you one of the main uses.

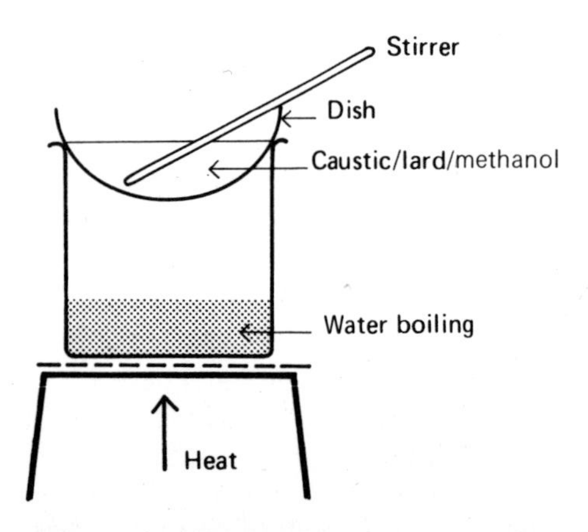

Fig. 26.8. Making some soap.

Put a piece of lard the size of a pea into an evaporating dish and add the same amount of caustic soda. Pour in 10 cm^3 of methanol using a measuring cylinder. Stir the mixture and warm it as shown in Fig. 26.8. This is a good way of warming things when they may catch fire. Keep stirring until the mixture goes into a paste. You have now made some soap, but it is very impure. Put a little of it into a test tube of water and then shake it. Dissolve some more and blow bubbles through it.

Making some colours

Blue: **$\frac{1}{4}$ fill a test tube with very dilute ferric (iron) chloride solution and add an equal amount of potassium ferrocyanide solution. Filter the blue chemical. Pour water through the substance while it is still on the filter paper – this is to wash it clean. When it has drained you can take the chemical out of the filter funnel whilst it is still in the paper, and put the cone in a warm place to dry. The name of the chemical is ferric (iron) ferrocyanide. You may know it under another name – Prussian Blue.**

Yellow: **Repeat the last experiment but use solutions of lead nitrate and potassium chromate. The chemical is called lead chromate.**

Red: **Repeat the experiment, but use solutions of potassium chromate and silver nitrate. The new chemical is called silver chromate.**

Now try mixing solutions of:

Potassium iodide and mercury chloride – what colour?

Copper sulphate and sodium hydroxide – what colour?

Barium chloride and sodium sulphate – what colour?

Which three elements are in lead nitrate? Sodium hydroxide? Potassium chromate? Sodium sulphate? Silver nitrate?

Which two are in Mercury chloride? Potassium iodide? Barium chloride?

Did any of these reactions give much heat? You *could* try one or two of them again with a thermometer in the test tube. These reactions are good fun to do because many of them give coloured chemicals. What do we call a solid substance that is formed when two solutions are mixed? Do you think a substance like this dissolves in water?

The chemicals you made are not soluble – we say *insoluble* – and this is why they make precipitates. You see how easy it is to make insoluble chemicals. Can you see what has happened during these reactions:

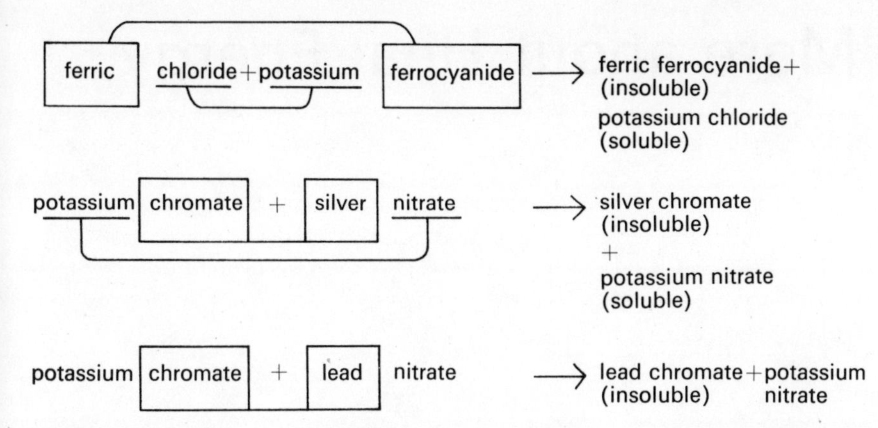

The metal part of one chemical combines with the acid part of the other. Which chemicals do we get with potassium iodide and mercury chloride, copper sulphate and sodium hydroxide, barium chloride and sodium sulphate? Fill in the blanks in your note-books:

(*a*) potassium............................ + barium nitrate ⟶ barium sulphate + potassium............................

(*b*) ....nitrate + sodium chloride ⟶ lead chloride + sodium nitrate.

(*c*) ....

Make your own indicator

Do you like pickled cabbage? Next time your mother makes some, do not let her throw the red cabbage water away. Put it in a pan and boil it gently so that some of the water evaporates and you are left with a strong red coloured liquid. See what this does in acids and alkalis.

More about hunting elements

Look back and find out the tests you did to find iron and chlorine and sulphur. You can now see how they work. What happens if chlorine is there and you add silver nitrate solution? What is the name of the precipitate? If you add silver nitrate to table salt which two chemicals will you get? We test for iron salts by adding potassium thiocyanate. What is the name of the red chemical which is made?

Summary

Acids are chemicals which have a sharp taste, they make soda give carbon dioxide and they can be detected best by using indicators. All of them have molecules containing hydrogen atoms.

When acids are used up by metals or soda or metal oxides, they leave a salt behind. We call metal oxides bases. When a chemical can use up acids and is also soluble in water we call it an alkali.

There are many salts and we call them after the acid part of them and the metal part.

Some salts are very easy to make because they are insoluble in water. When we mix the correct solution together the salt makes a precipitate.

Questions

(1) Write the names of six acids and four alkalis.
(2) Write the names of four bases and describe what they look like.
(3) Give the names of four insoluble salts.
(4) Which two chemicals are made when the following react together:
(*a*) nickel and hydrochloric acid
(*b*) magnesium and sulphuric acid
(*c*) zinc and sulphuric acid
(*d*) copper oxide and nitric acid
(*e*) aluminium oxide and hydrochloric acid
(*f*) caustic soda and citric acid?
(5) Give the names of two pairs of chemicals which could be used to make:
(*a*) magnesium chloride
(*b*) zinc sulphate
(*c*) copper nitrate
(6) Liquid A turned litmus red and gave a white precipitate with silver nitrate solution. Liquid B turned litmus blue and felt soapy. B also gave a yellow flame when a little of it was held in a Bunsen flame. What are A and B? What is made when A and B are mixed together?
(7) Lead chloride is insoluble in water. Which two chemicals would you use to make it and how would you do the experiment?
(8) Calcium sulphate is insoluble in water. Describe how you would make it.
(9) Write seven lines on each of these acids: lactic, carbolic, butyric, stearic, phosphoric, prussic. Show how we get them and what their uses are.

Chapter 27 More about Heat Energy

We have done some experiments in which we heated water and we have learned how to find out how many joules we need to warm up a certain mass of water through a certain number of degrees. Of course, we sometimes heat other things besides water, and so the next thing to find out is whether other substances are as easily heated as water, or whether it is easier to heat them, or more difficult.

Option A

We could of course compare them by putting equal weights in turn over a small flame, as we did with water, and seeing which got warmed up most in the same time. This would not be very safe for substances like paraffin that might catch fire. Can you think of a way in which you can heat paraffin without using a flame?

In the Experiment on page 112 we added hot water to an equal mass of cold water and we found when we did this the hot water lost some heat energy, the cold water gained some, and that the two quantities of energy were nearly equal. We also found that the temperature of the mixture was just about half-way between those of the hot and cold water.

To find out whether paraffin is more easily heated than water we can do an experiment like that one. Suppose we mix 100 g of hot water with 100 g of cold *paraffin*? If heating paraffin takes the same energy as heating water, what would you expect the final temperature to be? But suppose the paraffin is *harder* to heat than water. Would you expect the mixture to finish above the half-way temperature or below it? If the paraffin is easier to heat, what would you expect? We can safely try this experiment and find out. But this time we will try to keep as much heat as possible in the calorimeter.

Experiment: *To find out whether paraffin is easier or harder to heat than water.*

Fig. 27.1.

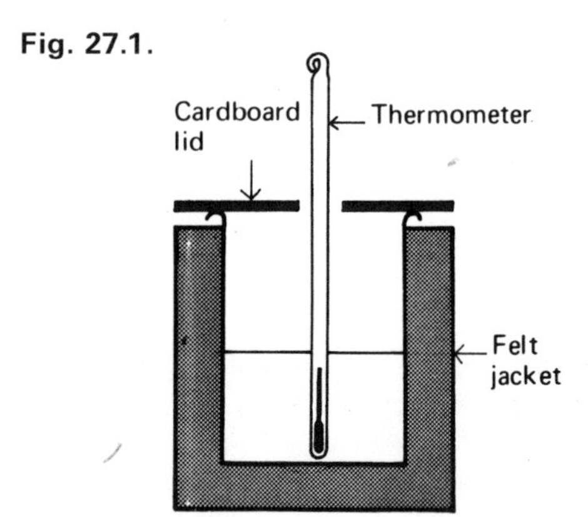

Look back to page 112 and see how the experiment on it was done. This time we are going to pour the hot water into cold paraffin instead of into cold water. You will use, say, 100 g of hot water and 100 g of cold paraffin. The density of paraffin is 0·8 g/cm³. How many cm³ of paraffin will you need? Use a measuring cylinder to pour the paraffin into the calorimeter, and then carry on as you did before (page 112), making a table of your observations as shown below. When you have finished the experiment, pour the mixture of paraffin and water into the vessel provided, so that the paraffin can be separated from the water and used again.

What was the temperature half-way between that of the hot and cold water? Was the final temperature of the mixture above this or below it? Do you think that the difference was due to heating of the calorimeter, or heating by the thermometer, or do you think that it is much too big to be caused in this way? If it *is* too big for this, then the difference must be because paraffin is more easy or more difficult to warm than water.

Observations:
Temp. of cold paraffin= °C
Temp. of hot water= °C
Temp. of mixture= °C
Rise in temp. of paraffin= °C
Fall in temp. of water= °C
Mass of paraffin used= g
Mass of water used= g

Remembering that 1 g of water loses 1 cal. (4·2 joules) when it cools 1° C, we can say,
No. of joules lost by hot water=

If all this heat was used to warm up the paraffin, we can say:

........g of paraffin are warmed .....° C by .... joules.
So 1 g of paraffin are warmed 1° C by .... joules.

This is not actually quite true, so our answer is only an *estimate* **(see page 20). Write down anything else that was warmed up besides the cold paraffin. We should have to allow for this if we wanted to get a really good answer and you will later learn how this is done.**

Why was a felt jacket put round the calorimeter, and why was a cardboard lid used?

Option B

Experiment: *To see whether different substances are equally easy to heat*

Fig. 27.2.

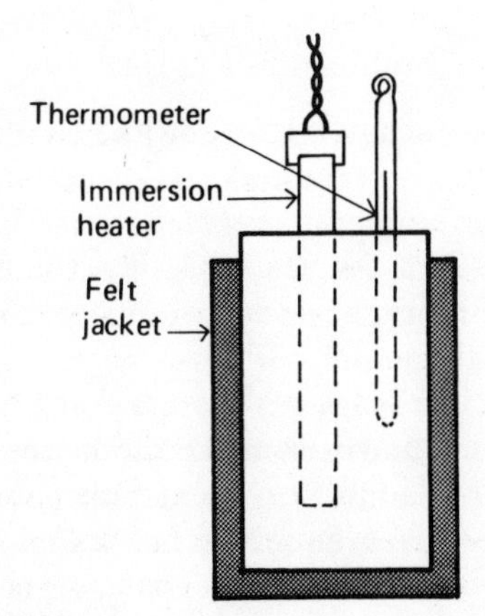

The metal blocks you are given each weigh one kilogramme, and you see that they have two holes drilled in them, one for the immersion heater and one for the thermometer. Why is a jacket provided? The heaters you use are all alike, so that all the blocks are given energy at the same rate. You may have been given a tin in which to heat water or paraffin. The density of water is 1 g/cm³, and of paraffin is 0·8 g/cm³. How many cm³ of water? of paraffin? must you measure out to get 1 kilogramme of each? Put this volume in your tin. Put the thermometer into the block or tin, leave it for two minutes and record the temperature in your book as follows:

Observations:
First temp. = °C
Second temp.= °C
Rise in temp. = °C
Heating time = min

When you are all ready you will be told to switch on the current, and you will switch it off again after a suitable time, which you will be told. As soon as you have switched off, read your thermometer and record your value.

Have all the thermometers risen by the same numbers of degrees? Are they *nearly* **the same? – if so it could be that the heaters weren't exactly alike, or it could be that some people hadn't read between the divisions (did you?). If the rises are very different it must be that some substances are more easy to heat than others. Which of the substances you used was the most difficult to heat? Which was the easiest? Make a table showing how many degrees you heated the various substances, thus:**

Temp. of copper rose °C
iron °C
water °C
paraffin °C
and so on.

The amount of heat energy needed to warm 1 g of a substance 1 deg C is called its *specific heat capacity*. If we know this about a substance we can work out how much heat is needed to warm up the substance, just as on page 111 we worked out how much heat we needed to warm different masses of water.

Example 1: How much heat is needed to warm 1 kg of copper from 15°C to 100°C?

We see from the information at the end that the specific heat capacity of copper is 0·37 J/g° C.

The copper is warmed from 15° C to 100° C, or (100 – 15)° C.

To warm 1 g of copper 1° C takes 0·37 J.
So to warm 1000 g of copper 1° C takes $0{\cdot}37 \times 1000 = 370$ J.
So to warm 1000 g of copper 85° C takes $85 \times 370 = 31\,450$ J.

Example 2: How much heat is given out when 500 g of silver cools from 90° C to 10° C?

The silver has cooled (90 – 10) = 80° C.

From the information table the specific heat capacity of silver is 0·24 joule per gramme per deg C.

When 1 g of silver cools 1° C it gives out 0·24 joule
So when 500 g of silver cools 1° C it gives out
$500 \times 0{\cdot}24 = 120$ joules
So when 500 g of silver cools 80° C it gives out
$80 \times 120 = 9600$ joules

Questions

(Use the information table at the end to help you work out the answers).

(1) How much heat is needed to warm 50 g of aluminium from 50° C to 100° C?
(2) How much heat is needed to warm 20 g of gold from 17° C to 87° C?
(3) How much heat is needed to warm 100 g of magnesium from 0° C to 70° C?
(4) How much heat is needed to warm 200 g of mercury from 40° C to 70° C?
(5) How much heat is needed to warm 80 g of tin from 12° C to 122° C?
(6) How much heat is given out when 5 kg of copper cools from 100° C to 20° C?
(7) How much heat is given out when 10 g of hydrogen cools from 100° C to 15° C?
(8) How much heat is given out when 1 kg of chloroform cools from 30° C to 5° C?
(9) How much heat is given out when 20 g of silver cools from 90° C to −30° C?
(10) How much heat is given out when 800 g of alcohol cools from 77° C to −13° C?

Information table

To warm 1 g of 1° C we need

copper	0·37 joule	or 0·09 cal
aluminium	0·84 joule	0·20 cal
gold	0·13 joule	0·03 cal
magnesium	1·1 joule	0·25 cal
mercury	0·12 joule	0·033 cal
tin	0·23 joule	0·055 cal
silver	0·24 joule	0·056 cal
hydrogen	14 joule	3·4 cal
chloroform	0·63 joule	0·15 cal
alcohol	1·9 joule	0·45 cal

In everyday life there are some interesting and important effects caused by the fact that different substances warm up different amounts for the same quantity of heat given to them. If you look at the table you will see that nearly all the substances are more easy to heat up than water is. This is true of the rocks and soil that make up dry land on the earth's surface. The surface of the earth is covered partly by land and partly by water (is there more land or more water?) Radiation from the sun falls on both land and water equally strongly. Where will the temperature rise most on a summer day – on land, or at sea? When the sun sets and the earth loses heat, where will it get coolest – on land or at sea? You may have noticed that on a hot afternoon at the seaside there is nearly always a gentle cool breeze blowing in from the sea. The diagrams in Fig. 27.3 show why. By early afternoon the land is much warmer than the sea, so a convection current of air occurs.

Fig. 27.3.

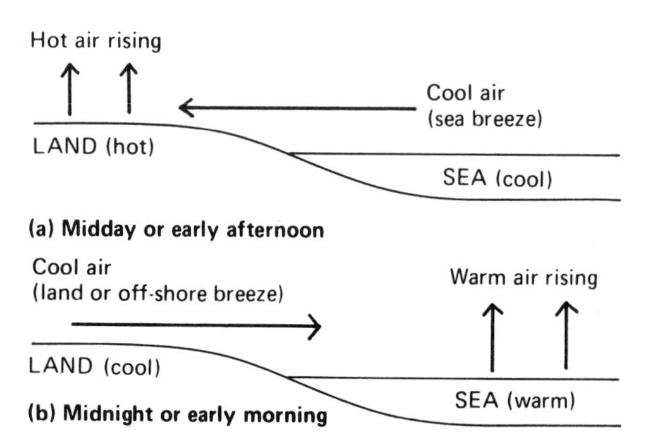

The air is warmed by the land and cooler air comes inland from the sea to take its place. On the other hand, in the early hours of the morning the land may be cooler than the sea, and then the convection current works the other way and a cool breeze blows off the land.

Because the sea heats up and cools down more slowly than the land, islands in the oceans and places on their coasts have cooler summers and warmer winters than places far inland. So the Canary Islands, for instance, off the African coast, never get as hot as the Sahara desert does inland.

The biggest land mass in the world is the continent of Asia, and during the summer the air over Asia becomes heated and rises, so that an enormous convection current is set up. The result is that cool air from the Pacific and Indian Oceans blows in towards the shores of Asia and brings to the people who live there the rain that enables them to grow their crops. These winds are called the Monsoons. So you see that to millions of people convection currents are really a matter of life and death.

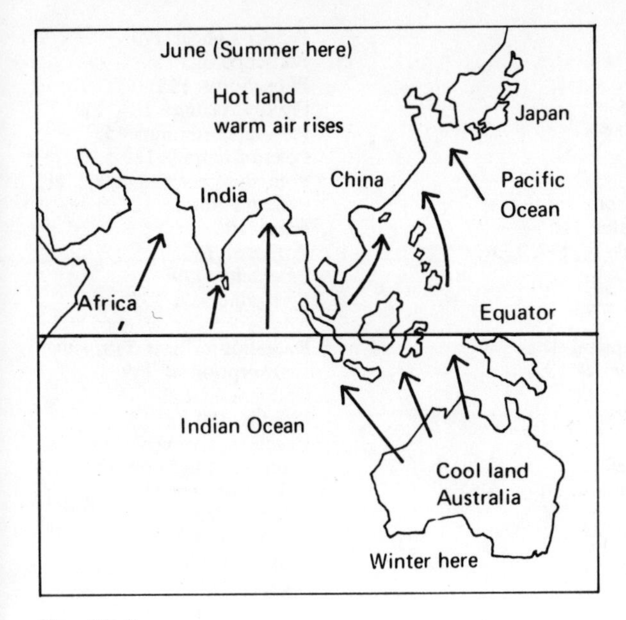

Fig. 27.4.

In winter the continent of Asia becomes very, very cold; this means that cold winds blow out from the land towards the oceans. Sometimes in winter in Britain we feel them blowing from the east and when that happens we know that we can expect some very cold weather. (See Figure 27.4).

What comes next?

During the next few years you will go on learning more science at school. There is still a lot of very interesting work to do for everybody. Some of you will go to work in laboratories when you leave school; all of you will meet scientists and read about their work in newspapers and hear of their work on television and radio.

We hope that you will go on finding things out for yourself by using your local library, doing experiments at home and checking up on ideas which seem doubtful. We have tried to show you that almost everything you come across in daily life is interesting to a scientist.

Most of all we hope that you have enjoyed science so far and that you will go on enjoying it.

Index